Materials Science and Technologies

MICROSTRUCTURALLY SHORT CRACKS IN POLYCRYSTALS DESCRIBED BY CRYSTAL PLASTICITY

MATERIALS SCIENCE AND TECHNOLOGIES

Magnetic Properties of Solids
Kenneth B. Tamayo (Editor)
2009. 978-1-60741-550-3

Building Materials: Properties, Performance, and Applications
Donald N. Cornejo and Jason L. Haro (Editors)
2009. 978-1-60741-082-9

Concrete Materials: Properties, Performance, and Applications
Jeffrey Thomas Sentowski (Editor)
2009. 978-1-60741-250-2

Mesoporous Materials: Properties, Preparation and Applications
Lynn T. Burness (Editor)
2009. 978-1-60741-051-5

Physical Aging of Glasses: The VFT Approach
Jacques Rault (Author)
2009. 978-1-60741-316-5

Graphene and Graphite Materials
H. E. Chan (Editor)
2009. 978-1-60692-666-6

Photoionization of Polyvalent Ions
Doris Möncke and Doris Ehrt (Authors)
2009. 978-1-60741-071-3

Handbook of Zeolites: Structure, Properties, and Applications
T. W. Wong (Editor)
2009. 978-1-60741-046-1

Dielectric Materials: Introduction, Research, and Applications
Ram Naresh Prasad Choudhary and Sunanda Kumari Patri (Authors)
2009. 978-1-60741-039-3

Strength of Materials
Gustavo Mendes and Bruno Lago (Editors)
2009. 978-1-60741-500-8

Handbook of Photocatalysts: Preparation, Structure, and Applications
Geri K. Castello (Editor)
2009. 978-1-60876-210-1

Corrosion Protection: Processes, Management, and Technologies
Teodors Kalniņš and Vilhems Gulbis (Editors)
2009. 978-1-60741-837-5

Handbook on Borates: Chemistry, Production, and Applications
M.P. Chung (Editor)
2009. 978-1-60741-822-1

Composite Laminates: Properties, Performance, and Applications
Anders Doughett and Peder Asnarez (Editors)
2010. 978-1-60741-620-3

Microstructurally Short Cracks in Polycrystals Described by Crystal Plasticity
Leon Cizelj and Igor Simonovski (Authors)
2010. 978-1-61668-811-0

Organometallic Compounds: Preparation, Structure, and Properties
H.F. Chin (Editor)
2010. 978-1-60741-917-4

Metamaterials: Classes, Properties, and Applications
Ethan J. Tremblay (Editor)
2010. 978-1-61668-958-2

Shape Memory Alloys: Manufacture, Properties, and Applications
H. R. Chen (Editor)
2010. 978-1-60741-789-7

Handbook of Material Science Research
Charles René and Eugene Turcotte (Editors)
2010. 978-1-60741-798-9

Organosilanes: Properties, Performance, and Applications
Elias B. Wyman and Mathis C. Skief (Editors)
2010. 978-1-60876-452-5

Europium: Compounds, Production, and Applications
Lucía M. Moreno (Editor)
2010. 978-1-61668-993-3

Smart Polymer Materials for Biomedical Applications
Songjun Li, Ashutosh Tiwari, Mani Prabaharan, and Santosh Aryal (Editors)
2010. 978-1-60876-192-0

Titanium Alloys: Preparation, Properties, and Applications
Pedro N. Sanchez (Editor)
2010. 978-1-60876-151-7

Piezoelectric Materials: Structure, Properties, and Applications
Wesley G. Nelson (Editor)
2010. 978-1-60876-272-9

Surface Modified Biochemical Titanium Alloys
Aravind Vadiraj and M. Kamaraj (Authors)
2010. 978-1-60876-581-2

High Performance Coatings for Automotive and Aerospace Industries
Abdel Salam Hamdy Makhlouf (Editor)
2010. 978-1-60876-579-9

Definition of Constants for Piezoceramic Materials
*Vladimir A. Akopyan, Arkady Soloviev, Ivan A. Parinov,
and Sergey N. Shevtsov (Authors)*
2010. 978-1-60876-350-4

Piezoceramic Materials and Devices
Ivan A. Parinov (Editor)
2010. 978-1-60876-459-4

Post-Impact Fatigue Behavior of Composite Laminates: Current and Novel Technologies for Enhanced Damage Tolernace
Alkis Paipetis and Dionysios T. G. Katerelos (Authors)
2010. 978-1-61668-672-7

Innovative Materials for Automotive Industry
Akira Okada (Author)
2010. 978-1-61668-237-8

Fundamentals and Engineering of Severe Plastic Deformation
*Vladimir M. Segal, Irene J. Beyerlein, Carlos N. Tome,
and Vladimir I. Kopylov (Authors)*
2010. 978-1-61668-190-6

Amines Grafted Cellulose Materials
Nadege Follain (Author)
2010. 978-1-61668-196-8

Lanthanide–Doped Lead Borate Glasses for Optical Applications
Joanna Pisarska and Wojciech A. Pisarski (Authors)
2010. 978-1-61668-292-7

Piezoelectric Ceramic Materials: Processing, Properties, Characterization, and Applications
Xinhua Zhu (Author)
2010. 978-1-61668-418-1

Recent Advances in Non-Destructive Inspection
Carosena Meola (Editor)
2010. 978-1-61668-550-8

Recent Developments in Silicone-Based Materials
Maria Cazacu (Editor)
2010. 978-1-61668-624-6

New Developments in Materials Science
Ekaterine Chikoidze and Tamar Tchelidze (Editors)
2010. 978-1-61668-852-3

Erbium: Compounds, Production, and Applications
Emily K. Byrne (Editor)
2010. 978-1-61668-989-6

Materials Science and Technologies

MICROSTRUCTURALLY SHORT CRACKS IN POLYCRYSTALS DESCRIBED BY CRYSTAL PLASTICITY

LEON CIZELJ
AND
IGOR SIMONOVSKI

Novika
Nova Science Publishers, Inc.
New York

Copyright © 2010 by Nova Science Publishers, Inc.

All rights reserved. No part of this book may be reproduced, stored in a retrieval system or transmitted in any form or by any means: electronic, electrostatic, magnetic, tape, mechanical photocopying, recording or otherwise without the written permission of the Publisher.

For permission to use material from this book please contact us:
Telephone 631-231-7269; Fax 631-231-8175
Web Site: http://www.novapublishers.com

NOTICE TO THE READER

The Publisher has taken reasonable care in the preparation of this book, but makes no expressed or implied warranty of any kind and assumes no responsibility for any errors or omissions. No liability is assumed for incidental or consequential damages in connection with or arising out of information contained in this book. The Publisher shall not be liable for any special, consequential, or exemplary damages resulting, in whole or in part, from the readers' use of, or reliance upon, this material.

Independent verification should be sought for any data, advice or recommendations contained in this book. In addition, no responsibility is assumed by the publisher for any injury and/or damage to persons or property arising from any methods, products, instructions, ideas or otherwise contained in this publication.

This publication is designed to provide accurate and authoritative information with regard to the subject matter covered herein. It is sold with the clear understanding that the Publisher is not engaged in rendering legal or any other professional services. If legal or any other expert assistance is required, the services of a competent person should be sought. FROM A DECLARATION OF PARTICIPANTS JOINTLY ADOPTED BY A COMMITTEE OF THE AMERICAN BAR ASSOCIATION AND A COMMITTEE OF PUBLISHERS.

Library of Congress Cataloging-in-Publication Data

Cizelj, Leon.
 Microstructurally short cracks in polycrystalls described by crystal plasticity / Leon Cizelj and Igor Simonovski.
    p. cm.
  Includes index.
  ISBN 978-1-61668-811-0 (softcover)
 1. Polycrystals--Plastic properties. 2. Polycrystals--Fatigue. 3. Microstructure--Mathematical models. 4. Grain boundaries--Mathematical models. I. Simonovski, Igor. II. Title.
  TA418.9.C7C56 2009
  620.1'126--dc22
                                                              2010016710

Published by Nova Science Publishers, Inc. † New York

Contents

Preface	xi
Nomenclature	1
Chapter 1 Introduction	3
Chapter 2 Computational Model	7
Chapter 3 Results	23
Chapter 4 Outlook	57
Chapter 5 Conclusion	63
Acknowledgments	65
References	67
Index	73

PREFACE

Microstructurally short cracks with lengths up to about ten grains are known to be strongly influenced by the microstructural features in the vicinity of a crack tip. These features include randomly shaped oriented crystal grains, and strong orientation dependent deformation behavior of the grains. The goal of our work is to propose computational models aiming to quantify the effects of random grain orientations on the variability of crack tip openings and sliding displacements (CTOD, CTSD).

A Voronoi tessellation based computational model has been developed to simulate the random grain structure. The constitutive behavior of individual grains includes randomly oriented anisotropic elasticity and crystal plasticity (where Schmid resolved stress is taken into account). The equilibrium equations are solved with macroscopic boundary conditions at the scale of the component using commercially available finite element solver ABAQUS.

The stationary crack configurations studied include transgranular crack extending through about half of a crystal grain, a series of cracks of different sizes simulating the short crack approaching and crossing the first grain boundary; and a series of cracks with different lengths extending from one to over a few grains. The FCC material with properties representing industry grade austenitic stainless steel is assumed with macroscopic uniaxial loading approaching macroscopic yield strength of the material. Sufficiently, many simulations with different random grain orientations have been performed to arrive at approximate cumulative probability distributions of the CTOD and CTSD. Possible limits of the CTOD/CTSD variability, as for example those derived from a large monocrystal with variable lattice orientations with respect to the crack and from the linear elastic fracture mechanics, are given and discussed. Discussion includes identification of at least two main sources of the CTOD/CTSD variability: grain structure and strain localizations extending over the entire computational domain. Also, the attempt is made to quantify the decreasing influence of the grain structure with increasing crack length.

The current computational model is limited to an essentially planar model (plane strain) with lattice rotations around the out of plane axis only. This basically allows for limiting most of the slip to two active (inplane) slip systems. The reason for this is the computational intensity of the simulations. A limited amount of simulations with spatial lattice 3D material orientations has also been performed to quantify the consequences of planar approximation.

An outlook towards modeling of as-measured spatial microstructures and intergranular cracks is given.

NOMENCLATURE

C_{ijkl}	4th order stiffness tensor
D_{ijkl} 4th	order compliance tensor
D_{ij}	symmetric rate of stretching tensor
$\dot{F}_{ik}$	deformation gradient
F_{kj}^{-1}	rate of the deformation gradient
F_{ij}	deformation gradient tensor
L_{ij}	velocity gradient
$\hat{L}_{kl}^{p}$	plastic velocity gradient in non-rotated coordinate system
P1	slip plane (111) with slip directions Dir1 [10$\bar{1}$], Dir2 [$\bar{1}$10] and Dir3 [0$\bar{1}$1]
P2	slip plane ($\bar{1}$11) with slip directions Dir1 [101], Dir2 [01$\bar{1}$] and Dir3 [$\bar{1}\bar{1}$0]
P3	slip plane (1$\bar{1}$1) with slip directions Dir1 [10$\bar{1}$], Dir2 [110] and Dir3 [0$\bar{1}\bar{1}$]
P4	slip plane (11$\bar{1}$) with slip directions Dir1 [011], Dir2 [1$\bar{1}$0] and Dir3 [$\bar{1}$0$\bar{1}$]
Q_{ij}	Rotation tensor
$R_{p0.2}$	yield strength
V	volume of the polycrystalline aggregate
α	index of the slip system
α	grain's crystallographic orientation
$\dot{\gamma}^{(\alpha)}$	slipping rate in slip system α
γ_0	amount of slip after which the interaction between slip systems reaches the peak strength
ε_{ij}	2nd order strain tensor
θ	crack direction
$\mu_{ij}^{(\alpha)}$	Schmid factor

σ_{ij}	2nd order stress tensor
τ_{ij}	Kirchoff stress
τ_{ij}^{∇}	Jaumann rate of Kirchoff stress
τ_0	yield stress
$\tau^{(\alpha)}$	Schmid resolved shear stress
Ω_{ij}	antisymmetric rate of spin tensor
$\dot{a}^{(\alpha)}$	reference strain rate in slip system α
$f_{\alpha\beta}$	the magnitude of the strength of particular slip interaction
$\dot{g}^{(\alpha)}$	current strain hardened state in slip system α
$g^{(\alpha)}$	current strength in slip system α
h_0	initial hardening modulus
$h_{\alpha\alpha}$	self-hardening moduli
$h_{\alpha\beta}$	latent-hardening moduli
h_S	hardening modulus during easy glide
i, j, k, l	indices running from 1 to 3
$m_j^{(\alpha)}$	slip plane normal
n	strain rate sensitivity parameter
q	hardening factor
$s_i^{(\alpha)}$	slip direction
t	time
x_j	vector of coordinates
v_i	velocity vector
u_i	deformation vector
AISI	American Iron and Steel Institute
CTOD	Crack tip opening displacement
CTSD	Crack tip sliding displacement
ITER	International Thermonuclear Experimental Reactor
LEFM	linear elastic fracture mechanics
*	elastic part
p	plastic part
$\langle\varepsilon_{kl}\rangle$	macroscopic strain tensor
$\langle\sigma_{kl}\rangle$	macroscopic stress

Chapter 1

INTRODUCTION

Aging and damage in materials could play a significant role in the long term safe operation of complex industrial systems such as, for example nuclear power plants. Considerable knowledge about aging and damage has accumulated over the years and has been extensively and successfully used in design and operation. Some remaining issues are however calling for better explanation. These include initialization and propagation of microstructurally short cracks, which account for a rather significant proportion of the component's life time.

The behavior of microstructurally short fatigue cracks with lengths in the order of ten grains or less distinctively differs from that of long cracks. In particular, the average crack growth rate can be much higher for short cracks with equivalent crack tip loading as first discovered by Pearson [1]. Short cracks have also been reported to grow below threshold values for long cracks [2].

Growth rate and path of short fatigue cracks are strongly affected by microstructural features such as grain boundaries, crystallographic orientations, inclusions, voids, and material phases etc. ([3]− [5]). Short fatigue cracks are often initiated from persistent slip bands and propagate along the slip planes. The crack tip loading is therefore generally mixed-mode and the crack changes direction as it passes through grain boundaries resulting in a serrated crack profile ([6]− [9]). Experimental data suggests that the fatigue crack growth rate and crack tip opening displacements tend to decrease when such cracks approach a grain boundary ([5], [10], [11]) and grain boundaries can also temporarily block the plastic zone growth [12]. The difficulty in propagating slip across an interface may give rise to an incubation period that depends on the type of interface, e.g., high-angle grain boundary or interface with a second phase. Different crystallographic orientations of the grains may also increase, decrease or arrest the crack growth [13], [14]. Vašek et al. [15] for example observed that crack propagation rates may vary significantly for nominally identical cracks. Crack closure and high strains can also play a significant role but plasticity induced crack closure effects are generally smaller for short cracks ([3], [4]).

Various models have been proposed to model the behavior of short cracks. Vašek et al. [15] and Polák et al. [16] both found that the propagation rate of microstructurally short cracks in AISI 316L can be fitted to a power law for the plastic strain amplitude that is almost independent of the crack size. Haddad et al. [17] could correlate short crack propagation to experimental data fairly well by extending the physical crack length with the transition length, for short and long cracks defined by Kitagawa diagrams [2]. Grain boundary blocking models that assume that the slip band zone is blocked at a grain boundary until an effective stress intensity factor is attained have also been used successfully to model crack retardation and acceleration (e.g., [18]–[20]). Alternatively the blocking is assumed to be related to the difference in crystallographic orientations between grains [20]. Other crack blocking models use separate equations and simply state that a long crack equation should be adopted if the threshold stress intensity factor is reached [21].

In recent years, several attempts have been made to model the behavior of long cracks in single crystal [22]; short cracks in bicrystal [23]; and polycrystals ([24], − [26]) using crystal plasticity material models but basically without the explicit grain shape modeling. Models with two ([13], [27], [28]); and more explicitly (see Bennett and McDowell [27]) modeled grains can be found in literature assuming a rectangular grain shape. The misorientation between two grains was found to have a smaller influence on the crack tip displacements, crack propagation rates and crack tip plastic zone than assumed in grain blocking models and some experimental data.

These references however are limited to a rather simplified modeling of grain shapes. It seems reasonable to extend such models by incorporating the basic randomness of grain shapes and lattice orientations within a large aggregate. These can have a significant effect on the crack tip loadings as shown by Ballarini et al. [29], and Wang and Ballarini [30]; where average values and standard deviations of stress intensity factors were calculated under the conditions of linear elastic fracture mechanics. It has been shown that the crack tip loadings are insensitive to the number of crystals and their orientation as long as the crack tip is surrounded by at least ten grains [29].

Introduction

Advancement of the models beyond the linear elastic regime required implementation of more sophisticated constitutive models. Crystal plasticity seemed to be the appropriate choice and had already been implemented in the multiscale modeling framework with constitutive modeling at the grain scale and solution of the boundary value problem at the component scale [31]. The randomly shaped grains were modeled using Voronoi tessellation. Some interesting results of the multiscale modeling framework include the estimates of typical representative volume element sizes [32] and correlation lengths in a typical nuclear pressure vessel steels [33]. Dislocations have not been modeled explicitly, although it is acknowledged that they are important, especially near the grain boundaries ([34], [35]). The first applications of the multiscale modeling framework towards short cracks were published (see in [36] and [37]) and estimated the scatter of the J-integral values for an intergranular crack extending over single grain boundary and compared with the isotropic case.

Further work has been focused towards short transgranular cracks in the randomly shaped and oriented polycrystalline aggregate. Examples include the scatter of crack tip opening displacement (CTOD) and crack tip sliding displacement (CTSD) for a surface crack embedded in a single grain [38]; and a crack kinking across the first grain boundary ([39], [40]). Also, the scatter in CTODs and CTSDs of a set of cracks extending over several grains have been estimated to show the vanishing influence of the microstructural features with increasing crack length [41]. These analyses assumed a stationary short crack initiating at the free surface of a planar (columnar) polycrystal under plane strain and monotonic loads. The CTOD and CTSD values have been used as a measure of the crack tip loading since they have been reported to describe the short crack growth better than stress intensity factors [11]. CTOD and CTSD also efficiently describe the mixed mode crack tip loading inherent in the Stage I crack growth.

In this work, the results obtained for the short inclined and kinked cracks extending up to about seven grains ([38] − [41]) are revisited. The compilation of the results has been performed to facilitate the generalization of already published observations and conclusions. The main contributions therefore includes the outline of the computational models and the quantification of the effects imposed by the random grain shapes.By describing orientations on the variability of crack tip opening and sliding displacements (CTOD, CTSD) for short cracks with various lengths ranging from about one half to seven grains. Some results have been presented using cumulative probability functions, which could facilitate a fundamental insight into the range of possible values attained experimentally. The results obtained for planar columnar aggregates are broadened with preliminary spatial analyses of transgranular short cracks and with an outlook towards the simulation of spatial integranular short cracks.

The material properties used are typical for the AISI 316L austenitic stainless steel; although extensively used in nuclear industry (e.g., pipes), ithas also been selected for a number of components of the International Thermonuclear Experimental Reactor ITER (vacuum vessel and ports, blanket shield modules, thin walled tubes for the first wall, cooling manifolds, diverter body etc.) [42].

Chapter 2

COMPUTATIONAL MODEL

Description of the computational multiscale modeling framework with constitutive modeling at the grain scale, and solution of the boundary value problem at the component scale is given in this section. More details are available in Kovač [31]. The essential features of the computational multiscale modeling framework include:

- The random polycrystalline structure is represented by a Voronoi tessellation. Each cell is assumed to correspond to a crystal grain - a monocrystal.
- The constitutive model of randomly oriented monocrystals assumes anisotropic elasticity and crystal plasticity.
- The boundary value problem is solved using the finite element method at the component scale.
- The overall properties of the polycrystalline aggregate are obtained by homogenization procedure.
- The crack tip opening displacement CTOD and the crack tip sliding displacement CTSD are estimated from the deformed finite element mesh.

These features for multiscale modeling are described in more detail. References to publications containing further details are made where appropriate.

The computational model is currently limited to planar (plane strain) simulations. The main reason for this limitation is vast computational intensity. The scaling to spatial simulations in the future is however straightforward.

Some preliminary spatial simulations already performed confirm the plausibility of results obtained by planar simulations ([43], [44]).

VORONOI TESSELATION

A Voronoi tessellation represents a cell structure constructed from a Poisson point process by introducing planar cell walls perpendicular to lines connecting neighboring points (Figure 1). This results in a set of convex polygons/polyhedra embedding the points and their domains of attraction, which completely fill up the underlying space. A survey about mathematical foundations and a variety of applications in different fields of science can be found for example in Aurenhammer [45]; and Stoyan *et al.* [46]. All Voronoi tessellations were generated using the code VorTess [47].

MESHING OF PLANAR VORONOI TESSELLATIONS

Generation of a suitable finite element mesh of a large Voronoi tessellation represents a rather challenging task. An automatic meshing algorithm described by Weyer *et al.* [48] has therefore been used. Each grain is treated as a separate entity, which creates very useful framework for keeping track of the individual grains. An example would be the assignment of different random crystal lattice orientations to each grain.

One of the basic requirements for reliable finite element analysis is suitable shape of the finite elements in the mesh. The discretization of the crystal grains into triangles is straightforward. The numerical quality of triangular finite elements, however, is generally poor. Planar quadrilateral finite elements (8 noded, reduced integration, plane strain) are therefore consistently used in the present meshing algorithms (Figure 1).

Only a subset of all possible tessellations could be meshed with quadrilaterals with reasonable internal angles and aspect ratii between the shortest and longest line within a single polygon. The "meshable" tessellations are obtained by the trial-and-error method. The bias introduced in the analysis by selecting only "meshable" tessellations is judged to be small compared to error caused by the 2-D approximation of grain structure [49].

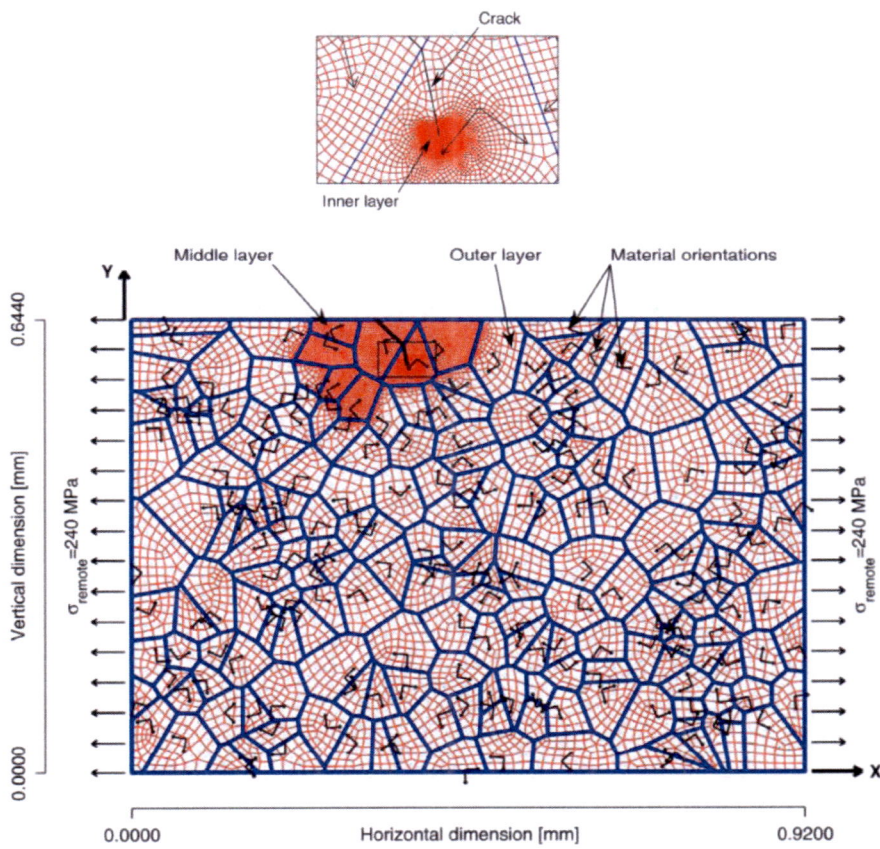

Figure 1. The outline of the finite element model including planar Voronoi tessellation, finite element mesh and random orientations of crystal lattices. (Reprinted from . Simonovski *et al.*, [38] © permission from Elsevier.)

LOADING AND BOUNDARY CONDITIONS

The applied macroscopic loading and boundary conditions are summarized in Figure 1. The left and right edges are loaded in macroscopic monotonic uniaxial tension from 208 MPa up to a maximum load of 280 MPa (83% to 112% of the macroscopic yield strength of 250 MPa) with zero shear traction. This load is sufficient to trigger significant slip systems activity in all cases analyzed. The macroscopic yielding is achieved in most cases analyzed. The upper and lower edges are traction free. Rigid body movement of the model is prevented.

CONSTITUTIVE MODEL

The essential components of the constitutive model are anisotropic elasticity, crystal plasticity, homogenization, and material parameterswill be described next.

Anisotropic Elasticity

Constitutive relations in linear elasticity are given by the generalized Hooke's law:

$$\sigma_{ij} = C_{ijkl} \cdot \varepsilon_{kl}, \qquad (1)$$

where σ_{ij} represents the 2nd order stress tensor, C_{ijkl} the 4th order stiffness tensor and ε_{ij} the 2nd order strain tensor. i, j, k and l are indices running from 1 to 3. In general, the stiffness tensor C_{ijkl} has 81 coefficients (elastic moduli). For the class of cubic crystals, the number of independent constants in stiffness tensor reduces to 3 in anisotropic and 2 in isotropic case [50].

The inverse of the stiffness tensor is called compliance tensor D_{ijkl} and is defined as:

$$\varepsilon_{ij} = D_{ijkl} \cdot \sigma_{kl}. \qquad (2)$$

Obviously, the compliance tensor D_{ijkl} of a cubic monocrystal is symmetrical and has again three independent coefficients.

Crystal Plasticity

The kinematics structure described in this section falls within the framework laid out by Taylor [51]; Rice [52]; Hill and Rice [53]; and Asaro and Rice [54], rigorously accommodating finite deformation effects. The plastic deformation is assumed due solely to the crystallographic slip [51]. The Schmid stress (or resolved shear stress on a slip system) is assumed to be the driving force [55], [56]). Deformation mechanisms such as diffusion, twinning and grain boundary sliding are not considered here. A user subroutine [57] incorporating the constitutive formulations outlined has been used in the finite element code ABAQUS.

Kinematics

The velocity gradient L_{ij} in the current state is given by a standard formula [56]:

$$L_{ij} = v_i \nabla_j = v_i \frac{\partial}{\partial x_j} = \frac{d}{dt}\frac{\partial u_i}{\partial x_j}, \tag{3}$$

where v_i represents velocity vector, x_j coordinate, u_i deformation vector and t time. The velocity gradient L_{ij} is computed in terms of the deformation gradient $\dot{F}_{ik}$ and its rate F_{kj}^{-1}:

$$L_{ij} = \dot{F}_{ik} \cdot F_{kj}^{-1} \tag{4}$$

The deformation gradient tensor F_{ij} is usually expressed as a product of the elastic part due to the lattice deformation with superscript * and the plastic part due to the plastic slip with superscript ^{p}, as illustrated in Figure 2:

$$F_{ij} = F_{ik}^* \cdot F_{kj}^p \tag{5}$$

Elastically deformed slip direction $s_i^{*(\alpha)}$ and slip plane $m_j^{*(\alpha)}$ for each slip system in crystal are given by [57]:

$$s_i^{*(\alpha)} = F_{ij}^* \cdot s_j^{(\alpha)} \quad \text{and} \tag{6}$$

$$m_j^{*(\alpha)} = m_j^{(\alpha)} \cdot F_{ij}^{*-1}. \qquad (7)$$

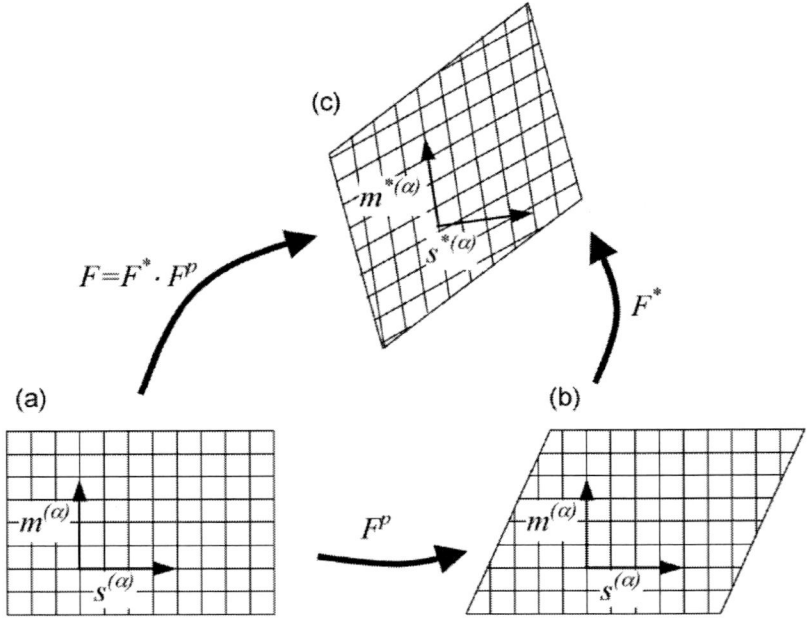

Figure 2. Kinematics of elasto-plastic deformation in crystalline materials.

The velocity gradient can be decomposed into elastic and plastic part:

$$L_{ij} = L_{ik}^* \cdot L_{kj}^p \qquad (8)$$

with elastic part L_{ij}^*:

$$L_{ij}^* = \dot{F}_{ij}^* \cdot F_{kj}^{*-1} \qquad (9)$$

and plastic part L_{ij}^p:

$$L_{ij}^p = F_{ik}^* \cdot \hat{L}_{kl}^p \cdot F_{lj}^{*-1} \qquad (10)$$

Plastic velocity gradient $\hat{L}^p_{kl}$ (in non-rotated coordinate system, indicated with (b) in Figure 2):

$$\hat{L}^p_{kl} = \dot{F}^p_{km} \cdot F^{p-1}_{ml} \qquad (11)$$

is given as the sum of the slips on all slip systems:

$$\hat{L}^p_{ij} = \sum_\alpha \dot{\gamma}^{(\alpha)} s_i^{(\alpha)} m_j^{(\alpha)}, \qquad (12)$$

where $s_i^{(\alpha)}$ and $m_j^{(\alpha)}$ are unit vectors in non-rotated coordinate system (Figure 2 (b)) defining the slip direction and slip plane, respectively and $\dot{\gamma}^{(\alpha)}$ is slipping rate. This can be transformed into:

$$L^p_{ij} = F^*_{ik} \cdot \left(\sum_\alpha \dot{\gamma}^{(\alpha)} s_i^{(\alpha)} m_j^{(\alpha)} \right) \cdot F^{*-1}_{lj}, \qquad (13)$$

$$L^p_{ij} = \sum_\alpha \dot{\gamma}^{(\alpha)} s_i^{*(\alpha)} m_j^{*(\alpha)}. \qquad (14)$$

Velocity gradient L_{ij} can also be written as a sum of stretching and spin part:

$$L_{ij} = D_{ij} + \Omega_{ij}, \qquad (15)$$

where D_{ij} and Ω_{ij} are the symmetric rate of stretching tensor and the antisymmetric rate of spin tensor, respectively. Both can be decomposed into lattice deformation and plastic parts:

$$D_{ij} = D^*_{ij} + D^p_{ij} \text{ and } \Omega_{ij} = \Omega^*_{ij} + \Omega^p_{ij}. \qquad (16)$$

Velocity gradient due to lattice deformation L_{ij}^* and plastic slip L_{ij}^p can also be decomposed into rate of stretching tensor and the rate of spin tensor:

$$L_{ij}^* = D_{ij}^* + \Omega_{ij}^* \text{ and } L_{ij}^p = D_{ij}^p + \Omega_{ij}^p, \tag{17}$$

with the plastic parts of stretching and spin rate tensors defined as:

$$D_{ij}^p = \frac{1}{2}\sum_{\alpha}\dot{\gamma}^{(\alpha)}\left(s_i^{(\alpha)}m_j^{(\alpha)} + s_j^{(\alpha)}m_i^{(\alpha)}\right)$$

and

$$\Omega_{ij}^p = \frac{1}{2}\sum_{\alpha}\dot{\gamma}^{(\alpha)}\left(s_i^{(\alpha)}m_j^{(\alpha)} - s_j^{(\alpha)}m_i^{(\alpha)}\right). \tag{18}$$

Constitutive Laws

We assume that the crystal elasticity is unaffected by slip. Then, the elastic law is defined as [56]:

$$\tau_{ij}^{\nabla*} = C_{ijkl} \cdot D_{kl}^* \tag{19}$$

where C_{ijkl} is the tensor of elastic modulus (non-rotating components, Figure 2 (b)) and $\tau_{ij}^{\nabla*}$ is Jaumann rate of Kirchoff stress in non-rotating coordinate system (Figure 2 (b)):

$$\tau_{ij}^{\nabla*} = \dot{\tau}_{ij} - \Omega_{ik}^*\tau_{kj} + \tau_{ik}\Omega_{kj}^* \tag{20}$$

The Kirchoff stress is defined as:

$$\tau_{ij} = \frac{\rho_0}{\rho}\sigma_{ij} \tag{21}$$

where σ_{ij} is Cauchy stress and ρ_0 and ρ are material densities in the reference and current state. τ_{ij}^{∇} is Jaumann rate of Kirchoff stress that rotates with the crystal lattice (Figure 2 (c)):

$$\tau_{ij}^{\nabla} = \dot{\tau}_{ij} - \Omega_{ik}\tau_{kj} + \tau_{ik}\Omega_{kj} \qquad (22)$$

The difference between these two rates is:

$$\tau_{ij}^{\nabla*} - \tau_{ij}^{\nabla} = \sum_{\alpha} \dot{\gamma}^{(\alpha)}\left(\omega_{ik}^{(\alpha)}\tau_{kj} - \tau_{ik}\omega_{kj}^{(\alpha)}\right), \qquad (23)$$

with tensor $\omega_{ij}^{(\alpha)}$ defined as:

$$\omega_{ij}^{(\alpha)} = \frac{1}{2}\left(s_i^{(\alpha)}m_j^{(\alpha)} - s_j^{(\alpha)}m_i^{(\alpha)}\right) \qquad (24)$$

Combining equations (14) and (23) results in the constitutive law in the form:

$$\tau_{ij}^{\nabla} = L_{ijkl} \cdot D_{kl} - \sum_{\alpha}\left(L_{ijkl}\mu_{kl}^{(\alpha)} + \omega_{ik}^{(\alpha)}\tau_{kj} - \tau_{ik}\omega_{kj}^{(\alpha)}\right) \cdot \dot{\gamma}^{(\alpha)} \qquad (25)$$

where Schmid factor $\mu_{ij}^{(\alpha)}$ is:

$$\mu_{ij}^{(\alpha)} = \frac{1}{2}\left(s_i^{(\alpha)}m_j^{(\alpha)} + s_j^{(\alpha)}m_i^{(\alpha)}\right). \qquad (26)$$

The Schmid stress is defined by:

$$\tau^{(\alpha)} = \mu_{ij}^{(\alpha)}\tau_{ij}, \qquad (27)$$

and the rate of change of Schmid stress by:

$$\dot{\tau}^{(\alpha)} = \left(\mu_{ij}^{(\alpha)}L_{ijkl} + \omega_{km}^{(\alpha)}\tau_{ml} - \tau_{km}\omega_{ml}^{(\alpha)}\right) \cdot D_{kl}^{*} \qquad (28)$$

$$\dot{\tau}^{(\alpha)} = m_i^{*(\alpha)} \cdot \left(\tau_{ij}^{\nabla*} - D_{ik}^{*}\tau_{kj} + \tau_{ik}D_{kj}^{*}\right) \cdot s_j^{*(\alpha)}. \qquad (29)$$

Hardening

The slipping rate $\dot{\gamma}^{(\alpha)}$ of the α-th slip system is determined by the corresponding resolved shear stress $\tau^{(\alpha)}$, based on Schmidt law and Hutchinson simple power law for polycrystalline creep:

$$\dot{\gamma}^{(\alpha)} = \dot{a}^{(\alpha)} \left(\frac{\tau^{(\alpha)}}{g^{(\alpha)}} \right) \left(\left| \frac{\tau^{(\alpha)}}{g^{(\alpha)}} \right| \right)^{n-1}, \qquad (30)$$

where $\dot{a}^{(\alpha)}$ is reference strain rate, n the strain rate sensitivity parameter, $\tau^{(\alpha)}$ the resolved shear stress and $g^{(\alpha)}$ the current strain hardened state of the crystal. In the limit as n approaches infinity this power law approaches that of a rate-independent material. The current strain hardened state $g^{(\alpha)}$ can be derived from:

$$\dot{g}^{(\alpha)} = \sum_{\beta} h_{\alpha\beta} \dot{\gamma}^{(\beta)}, \qquad (31)$$

where $h_{\alpha\beta}$ are the slip hardening moduli.

The present kinematic formulation of crystal plasticity constitutive model allows us to accommodate different hardening laws, which can be applied by supplying dependence of hardening moduli $h_{\alpha\beta}$ from the slip γ or it's derivates. One of the simplest hardening laws is Peirce et al. [58] and Asaro [56] hardening law. Self- (index $_{\alpha\alpha}$) and latent-hardening moduli ($_{\alpha\beta}$) are defined as:

$$h_{\alpha\alpha} = h(\gamma) = h_0 \operatorname{sech}^2 \left| \frac{h_0 \gamma}{\tau_s - \tau_0} \right|$$

and

$$h_{\alpha\beta} = q\, h(\gamma),\ (\alpha \neq \beta), \qquad (32)$$

where h_0 is the initial hardening modulus, τ_0 the yield stress, which equals the initial value of strength $g^{(\alpha)}(0)$, τ_s the break-through stress where large plastic flow initiates, q is hardening factor and γ cumulative slip on all slip systems integrated by time t:

$$\gamma = \sum_\alpha \int_0^t |\dot{\gamma}^{(\alpha)}| dt. \tag{33}$$

Wu et al. [59] and Bassani and Wu [60] have used different expression for the hardening moduli to describe the three-stage hardening of crystalline materials. Their expressions for self- (index $_{\alpha\alpha}$) and latent-hardening moduli ($_{\alpha\beta}$) depends on the shear strains of all systems:

$$h_{\alpha\alpha} = \left\{ (h_0 - h_S) \operatorname{sech}^2 \left| \frac{(h_0 - h_S)\gamma^{(\alpha)}}{\tau_S - \tau_0} \right| + h_S \right\} G(\gamma^{(\beta)}; \beta \neq \alpha) \tag{34}$$

$$h_{\alpha\beta} = q h_{\alpha\alpha} \quad (\alpha \neq \beta), \tag{35}$$

where the newly introduced h_S is the hardening modulus during easy glide within the stage I hardening. The function G is associated with latent hardening and given by:

$$G(\gamma^{(\beta)}; \beta \neq \alpha) = 1 + \sum_{\beta \neq \alpha} f_{\alpha\beta} \tanh\left(\frac{\gamma^{(\beta)}}{\gamma_0}\right). \tag{36}$$

γ_0 is the amount of slip after which the interaction between slip systems reaches the peak strength and $f_{\alpha\beta}$ represents the magnitude of the strength of particular slip interaction.

Rate-independent plasticity is treated here as the limit or rate dependent visco-plasticity [57].

Homogenization

The overall homogenized properties of the polycrystalline aggregate are an approximation of the properties observed in the macroscopic engineering world which basically assumes homogeneity of the materials. The macroscopic stress $\langle\sigma_{kl}\rangle$ and macroscopic strain tensors $\langle\varepsilon_{ij}\rangle$ are estimated using volume averaging [61]:

$$\langle\sigma_{kl}\rangle = \frac{1}{V}\int_V \sigma_{ij}\, Q_{ik}\, Q_{jl}\, dV,$$

$$\langle\varepsilon_{ij}\rangle = \frac{1}{V}\int_V \varepsilon_{ij}\, Q_{ik}\, Q_{jl}\, dV, \qquad (37a, b)$$

Where σ_{kl} and ε_{ij} stand for local stress and strain tensors, respectively, and V for volume of the polycrystalline aggregate. Rotation tensor Q_{ij} is illustrated in Figure 3.

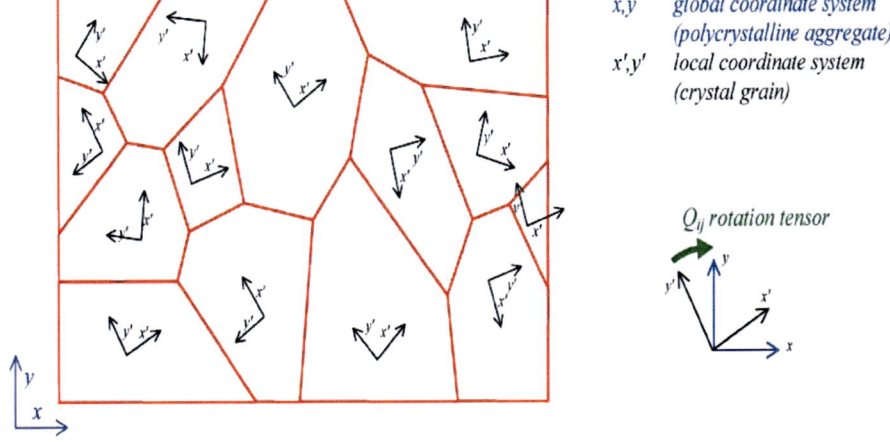

Figure 3. Global and local coordinate systems.

Material Parameters

The following elastic constants for the AISI 316L face centered cubic crystal are used: C_{iiii}= 163.680 GPa, C_{iijj}= 110.160 GPa, C_{ijij}= 100.960 GPa [62]. The average grain size used in the analysis is 52.9 μm.

Crystal plasticity parameters have been optimized from the macroscopic plastic response of AISI 316L polycrystal [62]: h_0= 330MPa, τ_s= 270MPa, τ_0= 90MPa, n= 55, q= 1.0 and $\dot{a}^{(\alpha)}$ = 0.001. With these parameters the proposed plain strain model is deemed sufficiently accurate to provide a correct qualitative representation of the macroscopic response [62]. The reference calculations with isotropically elastic and plastic material models have been carried out using Youngs modulus of 192 GPa, Poisson ratio of 0.3, yield strength of 250 MPa and hardening modulus of 1.118 GPa.

LAYOUT OF CRACKS

Orientations of Cracks Relative to Slip Planes

The orientation of lattice was defined in two basic steps:
1. The angle between the crystallographic [100] direction and the macroscopic X direction of all crystals in the model was fixed at 135°, as shown in Figure 4. A direct consequence of this orientation is a planar slip system model compatible with the planar macroscopic model. The projections of the primary and conjugate slip planes active in a planar model are depicted in Figure 5.
2. Since each grain is assumed to behave as a randomly oriented monocrystal, random rotation of the lattice in individual grains about the global Z-axis is imposed. The angle of this random rotation is hereafter referred to as crystallographic orientation α. (Figure 4 and Figure 5).

The crack is always placed in a slip plane. This requires that the orientation of the crack containing grains is fixed rather than chosen at random. The choice between the two active slip planes however remains. An arbitrary assumption has been made that the crack always initiates in the slip plane denoted as P2 (Figure 4 and Figure 5).

Increasing the crack length requires crossing the grain boundaries and kinking of the crack direction. Again, it is assumed that the crack follows the slip plane also in the newly damaged grain. The choice between two possible slip planes after the crossing of grain boundary has been elaborated based on parametric studies discussed here.

The direction of the crack θ, measured in the counter clockwise direction, will coincide with the slip plane P2 in the lattice with orientation α when (Figure 5):

$$\theta = 90° + 35.264° + \alpha + 180°. \tag{38}$$

The example depicted in Figure 5 assumes that the crack direction is $\theta=315°$. To align the crack with the slip plane P2, the lattice must be rotated by $\alpha=9.735°$.

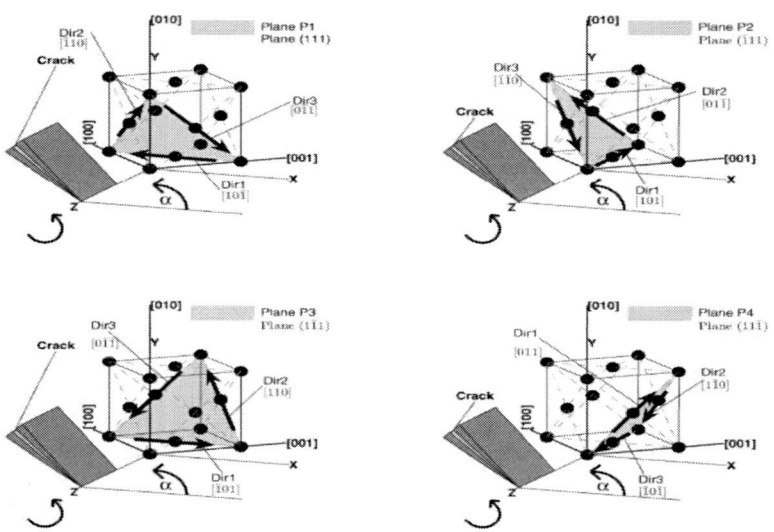

Figure 4. Orientation of crack with respect to computational plane and lattice. (Reprinted from Simonovski et al., [38] © permission from Blackwell Publishing).

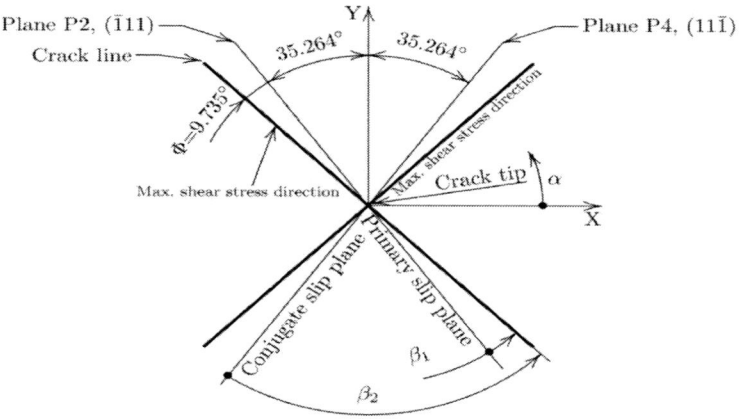

Figure 5. Orientations of slip planes and crack for crystallographic orientation $\alpha=0°$. (Reprinted from Simonovski et al., [38] © permission from Blackwell Publishing.)

Finite Element Mesh at the Crack Tip

The mesh is refined significantly in the vicinity of the crack tip, as depicted in Figure 1. As a rough guide, the typical element size away from the crack tip is 10 µm. This reduces to about 0.25 µm for the elements at the crack tip and to about 0.125 µm for the crack tips in the immediate vicinity of the grain boundary.

The average grain size is 52.9 µm. The values of the CTOD and CTSD have been extracted at the distance 1.3 µm from the crack tip (Figure 6). Such definition is consistent with examples found in the literature (e.g., [27], [28]). The mesh refinement adopted in the analyses presented here has been extensively tested and shown to provide reasonably reliable CTOD and CTSD estimates [63].

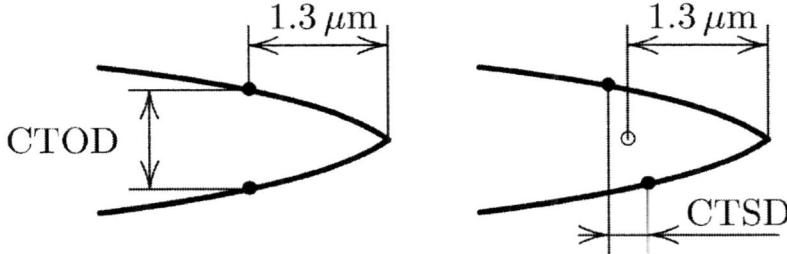

Figure 6. Definitions of the crack tip opening displacement CTOD and the crack tip sliding displacement CTSD. (Reprinted from Simonovski et al., © permission from Blackwell Publishing.)

Chapter 3

RESULTS

STATIONARY OBLIQUE SURFACE CRACK

The structural model is a planar rectangular aggregate with 212 randomly sized and shaped grains, as depicted in Figure 1. A short inclined surface crack with length 35.44 μm is introduced in the grain #38 at the top surface of the model. The size of the grain, estimated as the square root of its area, is equal to 70.87 μm, which represents double crack length. The macroscopic orientation of the crack is fixed to $\theta=135°$ relative to the global X axis. To place the crack into the primary slip plane, the crystallographic orientation of the cracked grain has been set to $\alpha = 9.735°$ (see Figure 5).

MONOCRYSTAL

Before embarking on the discussion of the results obtained with rather complex models containing randomly oriented crystal grains (monocrystals) it may be useful to give some insight in the constitutive response of anisotropic monocrystals. Changing the orientation of the lattice in the monocrystal with respect to the uniaxial load namely directly influences the stiffness of the monocrystal (individual grain) in the direction of the macroscopic loading (Figure 7, $\sigma_{11} = 240$ MPa). The finite element model depicted in Figure 1 was used for this purpose with the crystallographic orientations of all 212 grains set to identical value, varying from 0° up to 180° in 2° increments.

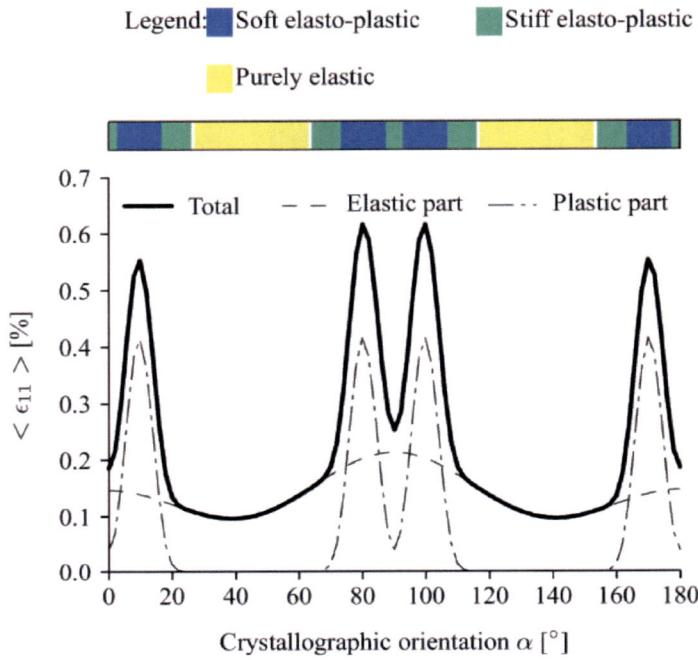

Figure 7. Macroscopic strain $<\varepsilon_{11}>$ in a monocrystal as a function of crystallographic orientation. (Reprinted from Cizelj et al., [41] © permission from American Society of Mechanical Engineers.)

The contribution from the anisotropic elasticity (dashed line in Figure 7) is notable, but moderate. The largest contribution comes from the crystal plasticity: first, the stiffness in crystal plasticity is defined by the Schmid factors eq. (26); and second, by the power law dependence of the slipping rate, eq. (30). Due to the rather high exponent in eq. (30), directions with slightly smaller Schmid factors will exhibit significantly smaller slipping rate. The combination of the two factors results in very distinct directions with stiff and soft response of the monocrystal.

Three somewhat arbitrary levels of the stiffness in the directional response of the monocrystal are defined here for illustrative purposes: (1) purely elastic response, (2) stiff elastic-plastic with $0 <= \varepsilon^{pl}_{11} < 0.1\%$, and (3) soft elastic-plastic response with $0.1\% <= \varepsilon^{pl}_{11}$ (Figure 7). We refer to this classification as 'grain hardness index.' This will come in handy later on in the polycrystal configuration to display soft and hard grains.

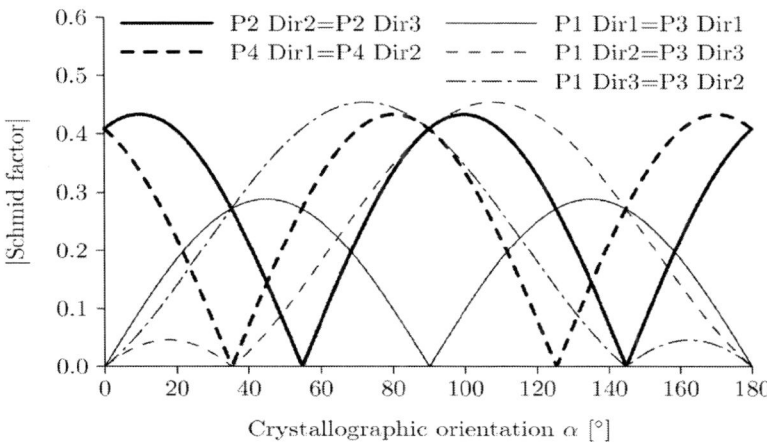

Figure 8. Schmid factors as a function of crystallographic orientation. (Reprinted from Simonovski *et al.*, [38] © permission from Blackwell Publishing.)

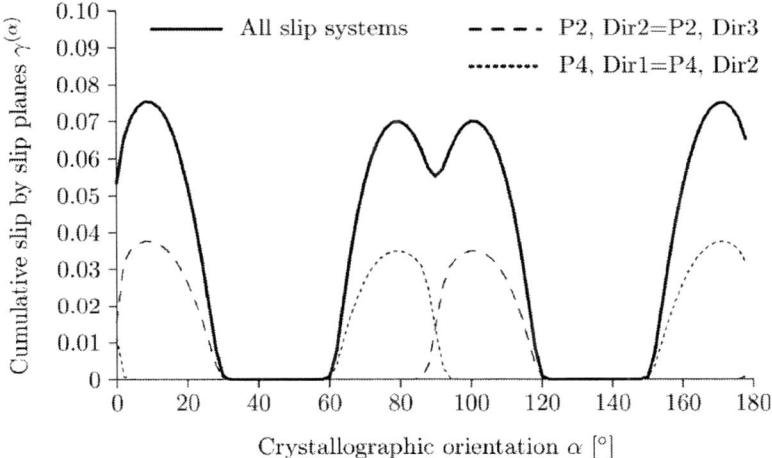

Figure 9. Cumulative slip in active slip systems as a function of crystallographic orientation far from the crack tip. (Reprinted from Simonovski *et al.*, [38] © permission from Blackwell Publishing.)

Schmid factors (eq. (26)) for all slip systems are depicted in Figure 8. Comparing these factors with Figure 7 one can, as expected, see that the maximal strains are obtained at crystallographic orientations where the Schmid factors on planes P2 (Dir2 and Dir3) and P4 (Dir1 and Dir2) are at their highest values as well. We will show later that, with the exception of the immediate vicinity of the crack tip, only these four slip systems are active in the model. Slip planes, directions and Miller indices and vectors are shown in Figure 4. The dominating crystal plasticity contribution to the macroscopic strain is clearly seen in Figure 7 at the orientations $0° \leq \alpha \leq 30°$, $60° \leq \alpha \leq 120°$ and $150° \leq \alpha \leq 180°$.

Cumulative slips (eq. (33)) in the active slip systems far away from the crack tip are depicted in Figure 9 and correlate very well with the macroscopic strains in Figure 7. As expected (Figure 5), four slip systems are active far away from the crack tip: P2 (Dir2 and Dir3) and P4 (Dir1 and Dir2). Activity of the rest of the slip systems is negligible. In the vicinity of the crack, more complex slip system activity is noted: slip systems P1, Dir1 and P3, Dir1 are activated in addition to the activated systems in elements far from the crack tip Figure 10.

CTOD and CTSD at the load levels of 240 MPa ($0.96 R_{p0.2}$) and 280 MPa ($1.12\ R_{p0.2}$) are depicted in Figure 11 as a function of crystallographic orientation. The correlation between low CTOD regions and small cumulative slip regions (Figure 10) is excellent. Furthermore, these regions coincide well with small $<\varepsilon_{11}>$ regions (Figure 7), and small Schmid factors (Figure 8). However, they are narrower. The largest crack tip displacements occur at crystallographic orientations where the Schmid factors are high. In these cases almost all part of a crack tip displacement is due to the plastic deformation.

The largest crack tip displacements are noted at the crystallographic orientation that would result in symmetric slip at $\alpha = 90°$ (note that the CTSDs are not zero since the crack is at an angle of 45°). This result is to some extent surprising, since we expected the CTOD and CTSD maxima for the crack aligned with the slip plane ($\alpha = 9.735°$ or $\alpha = 80.264°$). The most complex slip system activity isnoted for $\alpha = 90°$: active slip systems include P2 (Dir2 and Dir3) and P4 (Dir1 and Dir2) plus P1 Dir1 and P3 Dir1, all of them being close to the maximum Schmid factor (see Figure 8). Since the activation of the slip systems other than the in plane P2 (Dir2 and Dir3) and P4 (Dir1 and Dir2) is to a large extent triggered by the local strain field and by the plane strain constraint, a future 3-D analysis will be needed to clarify this issue. The extreme CTOD (CTSD) values recorded at remote load of 280 MPa (240 MPa) were 2.13μm (1.84μm) and 0.12μm (0.097μm), respectively.

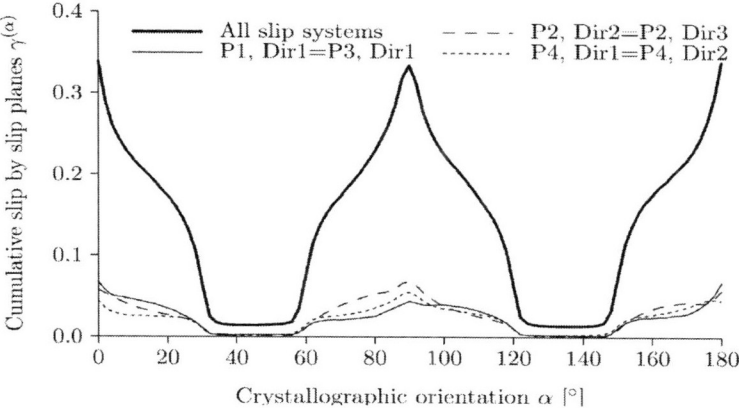

Figure 10. Cumulative slip in active slip systems as a function of crystallographic orientation near the crack tip. (Reprinted from Simonovski et al., [38] © permission from Blackwell Publishing.)

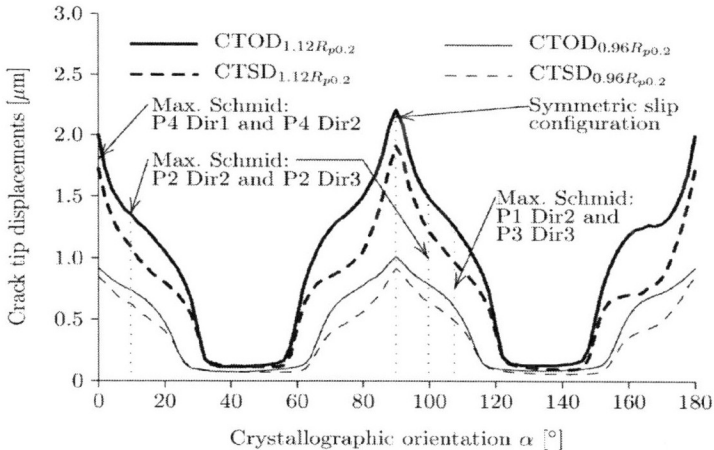

Figure 11. Crack tip displacements as a function of crystallographic orientation. (Reprinted from Simonovski et al., [38] © permission from Blackwell Publishing.)

The most unfavorable direction of the crack is, for the purpose of the analysis bellow and pending the 3-D resolution of local slip systems activity, assumed at $\theta=135°$. To align the crack with the slip plane, the crystallographic orientation must then be set to $\alpha = 9.735°$. At remote load of 280 MPa this orientation leads to CTOD and CTSD of 1.29μm and 1.04μm, respectively.

Bicrystal

Before embarking on the random polycrystalline case it was considered interesting to analyze also the behavior of a large bicrystal. Based on favorable and unfavorable configurations previously developed, a model of a bicrystal has been constructed as follows: the orientation of the crack containing grain was set to $\alpha = 9.735°$ (Stage I crack) while all the remaining grains were oriented at $\alpha = 135°$ to minimize CTOD as much as possible.

This resulted in the CTOD value of $0.134 \mu m$, which is moderately higher than obtained by orienting all the grains in the unfavorable orientation of $\alpha = 135°$ ($0.116 \mu m$). This indicates considerable importance of the orientation of the crack containing grain.

Polycrystal

The polycrystal case was simulated by applying 100 sets of random orientations (uniform distribution for 0 to 2π) to 211 out of the 212 grains in the model outlined in Figure 1. The orientation of the crack containing grain #38 was kept fixed to 9.735° in all simulated cases to mimic the Stage I crack aligned with the slip plane. The resulting cumulative probability of CTOD (CTSD) values is depicted in Figure 12 (Figure 13) and labeled as 9.735, random case. The computed CTOD values fall mainly between the extreme values obtained for the large monocrystal. The scatter of the results is completely attributed to the random orientations of the intact grains.

The initial orientations of grains for the case with maximum and minimum CTODs are plotted in Figure 14 and Figure 15, respectively, using the grain hardness index (see Figure 7). It is clear from Figure 14 and Figure 15 that the magnitude of the CTOD strongly depends on the grain clusters with soft elasto-plastic response leading to a localized soft response (e.g., shear band) of the polycrystal. If such a cluster develops in the vicinity of the crack tip, the localized strain becomes the major contributor to a large CTOD value (Figure 14). On the contrary, when a soft cluster is formed away from the crack tip, the localized strain does not affect the crack tip directly, which results in a small CTOD value (Figure 15).

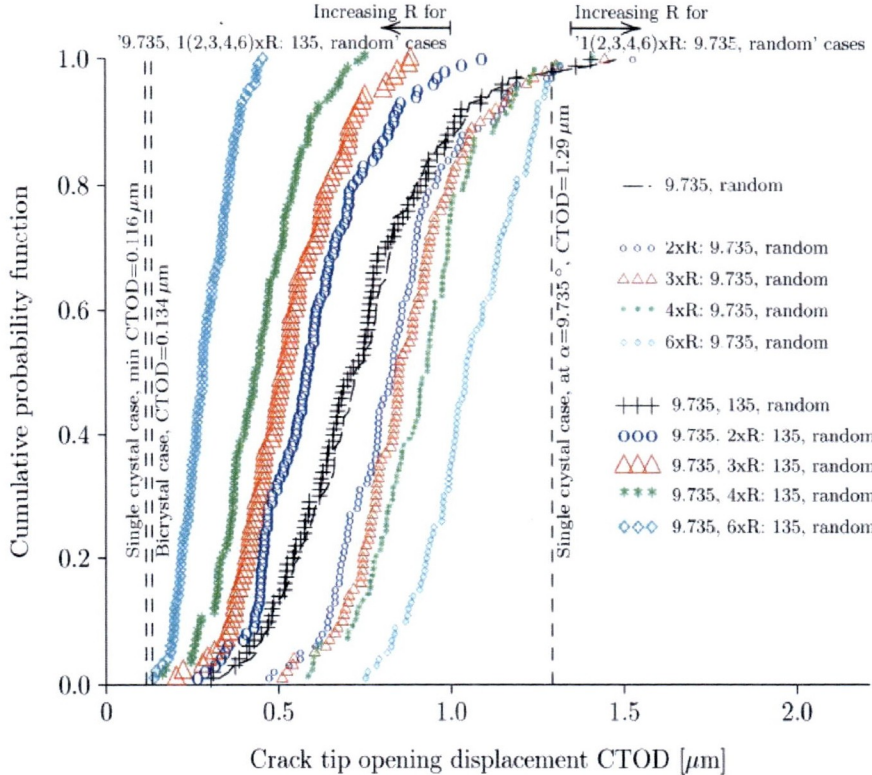

Figure 12. Cumulative probability functions for CTOD at remote load of 280 MPa. (Reprinted from Simonovski et al., [38] © permission from Blackwell Publishing.)

This indicates that the orientations of grains far away from the crack tip may contribute significantly to the development of localized strain patterns, which are significantly larger than the size of a typical grain. The position of localized strain patterns may in turn significantly influence the CTOD values.

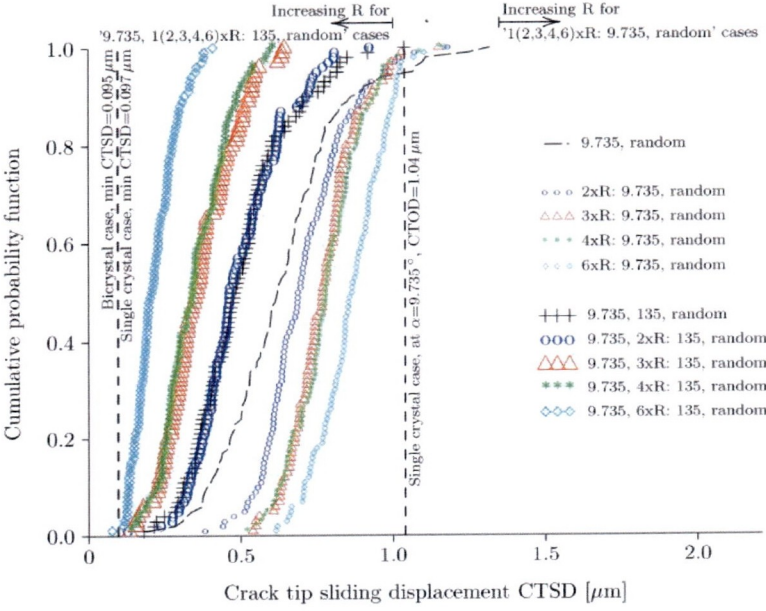

Figure 13. Cumulative probability functions for CTSD at remote load of 280 MPa.

Sensitivity to the Grain Size

In addition to the localized strain patterns, the relative size of the crack containing grain and the unfavorably oriented grain in its immediate vicinity are expected to have strong effect on the development of the CTOD and CTSD. Two sets of simulations have been performed to assess this effect:

1(2,3,4,6)xR: 9.735, random case mimics the increasing size of the grain containing the crack. This has been achieved by setting the orientation of the crack containing grain and all grains falling into radii of 2, 3, 4 and 6x crack lengths to $\alpha = 9.375°$. All remaining grains assumed random orientation (100 different realizations). A notation of 3xR: 9.735, random for example indicates that all grains having their Poisson points within the radius of three times crack length (3xR) are oriented at $\alpha = 9.375°$, while all other grains are oriented randomly. The radii are indicated in Figure 14 and Figure 15. Monocrystal size of 6xR is the maximal size considered to avoid significant effects from the limited size of the model. More grains would have to be employed to investigate this matter further in the future.

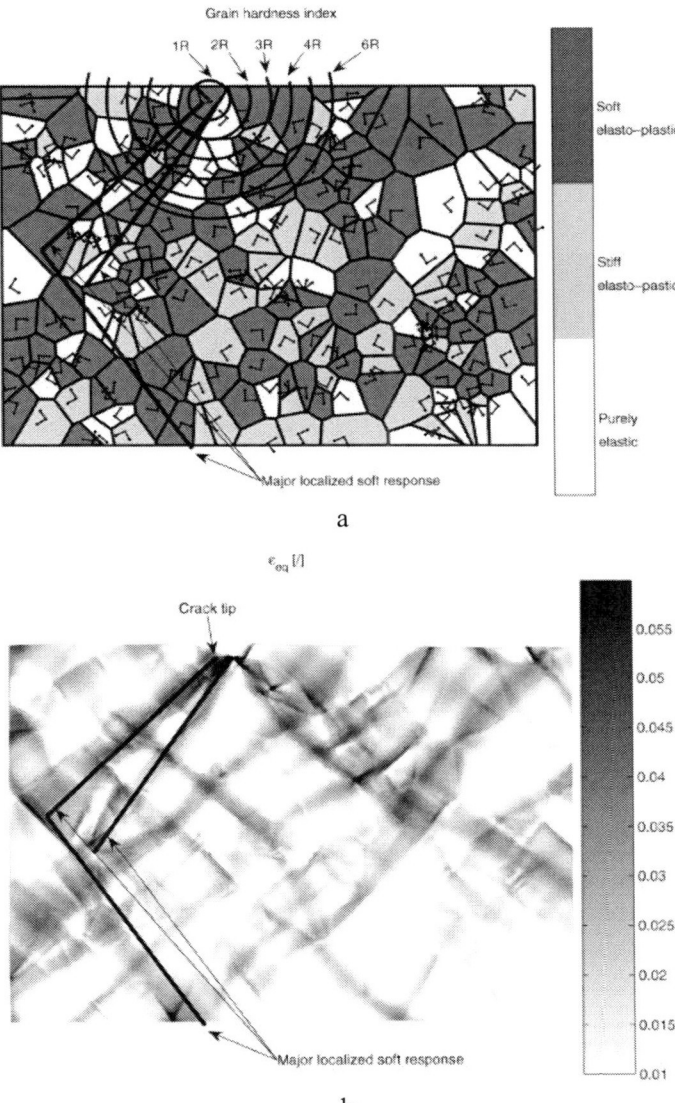

Figure 14. Grain hardness index (a) and equivalent strain (b) for the 9.735,135, random set of crystallographic orientations resulting in large CTOD. (Reprinted from Simonovski *et al.*, [38] © permission from Blackwell Publishing.)

9.735, 1(2,3,4,6)xR: 135, random case mimics the increasing size of the unfavorably oriented grain in the immediate vicinity of the crack containing grain. It is useful to understand these cases as a deformable, but increasingly strong clamp around the crack containing grain. The orientation of the crack containing grain was set to $\alpha = 9.375°$, the orientation of the first neighboring grain in the direction of the crack growth has been set to the unfavorable $\alpha = 135°$. The size of the unfavorably oriented grain has then been increased by applying the $\alpha = 135°$ orientation to all grains falling into radii of 2, 3, 4 and 6x crack lengths, with the obvious exception of the crack containing grain. All remaining grains assumed random orientation (100 different realizations). A notation of 9.735, 3xR: 135, random for example indicates that the orientation of the first grain is 9.735°, orientation of grains having their Poisson points within radius of three times crack length (3xR) is 135°, while all other grains are oriented randomly.

Figure 12 and Figure 13 show the cumulative probability functions for these cases. As expected, the results are mainly contained within the bounds posed by the monocrystal case, which are plotted for comparison. Also, the CTOD and CTSD values for a bicrystal case (crack-containing grain: $\alpha = 9.375°$, all other grains: $\alpha = 135°$) are shown. A small amount of values characterized as the top 5% exceeds the limits posed by the monocrystal. Those values are believed to be dominated by shear bands in the vicinity of the crack tip.

Min/max CTOD values for a limiting case of a single crystal are plotted for comparison. One can see that for the 9.735, random case the random crystallographic orientation can change the CTOD by a factor of 4.5 (c.f., Table 1). If we consider the values from all the analyzed cases, the ratio CTODmax/CTODmin increases to 11.

Increasing the size of the crack-containing monocrystal to 2, 3, 4, 6xR gradually increases both the average values of CTOD and CTSD. Their spread at the other hand tends to decrease. It is interesting to note that the impact of the increasing monocrystal size vanishes at large CTOD and CTSD values, where cumulative densities of 2, 3, 4 and 6xR overlap (top 5%). This indicates that in this region is dominated by the localized high strain areas (areas of grains with favorable orientation in the direction of maximum shear). It is interesting to note that increasing the size of the monocrystal could slightly reduce this effect on the CTSD.

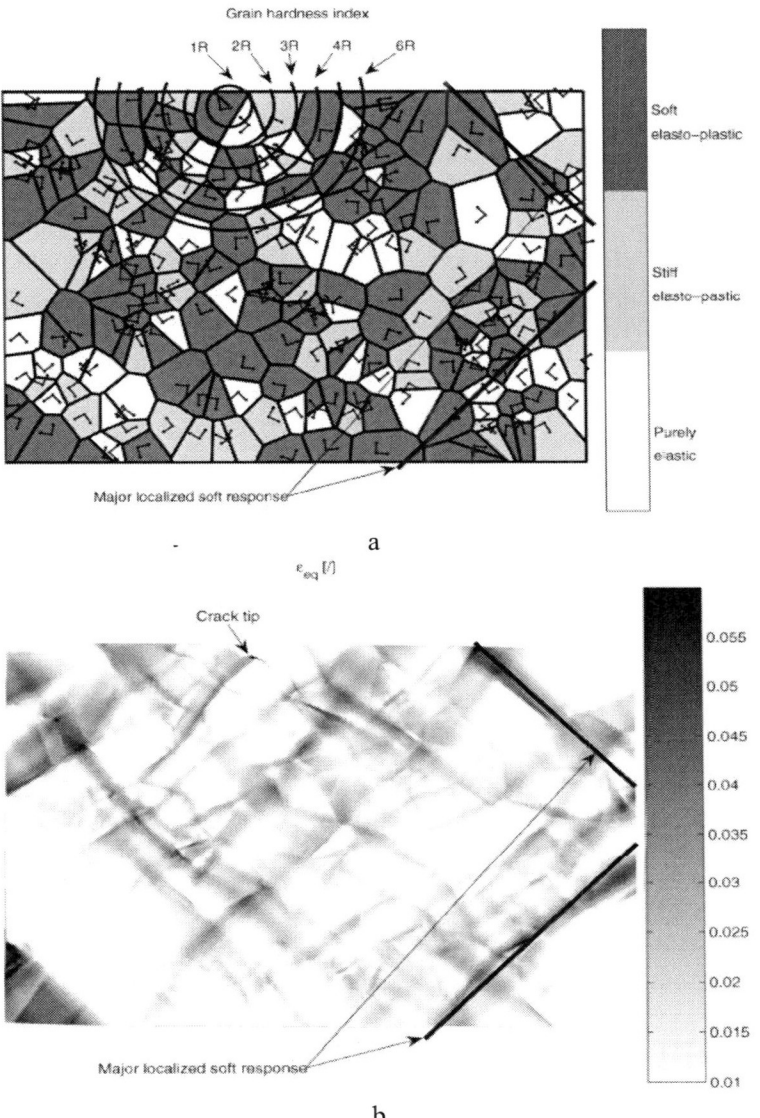

Figure 15. Grain hardness index (a) and equivalent strain (b) for the 9.735,135, random set of crystallographic orientations resulting in small CTOD. (Reprinted from Simonovski *et al.*, [38] © permission from Blackwell Publishing.)

It is also interesting to note that the largest CTOD and CTSD can actually be higher than predicted by the limiting case of a single crystal. We again attribute this to the occurrence of localized high strain areas in the polycrystal case.

As expected, increasing the size of the $\alpha = 135°$ layer (9.735, xR: 135, random cases), e.g., the clamp around the crack containing grain, decreases both the CTOD and CTSD values. The clear lower limit in this situation is the bicrystal case. The spread of the CTOD/CTSD values gradually decreases. However, the random structure beyond 6xR still significantly affects the CTOD and CTSD values. It is also noted that the bicrystal configuration successfully minimizes the effects of the localized high strain areas due to its high stiffness.

As a rule of the thumb, the values of CTSD are about 10% lower than CTOD in all analyzed configurations.

CRACK CROSSING THE FIRST GRAIN BOUNDARY

The oblique crack previously described and depicted in Figure 1 is now assumed to grow towards and across the first grain boundary. This has been mimicked by a set of stationary cracks with different lengths, as depicted in Figure 16. The dots on the crack denote the subsequent positions of the crack tips.

The direction of the crack ($\theta=135°$) and the orientation of the lattice in grain #38 ($\alpha = 9.735°$) are those assumed as CTOD maximizing for the oblique crack (section 0). As the crack tip crosses the grain boundary and enters the subsequent grain #124, it is again assumed to follow the slip plane.

It is useful to define the direction of the crack growth into the second grain with a relative measure: crack extension deflection $\Delta\theta_{124}$, as depicted in Figure 16. For the crack growing into the subsequent grain #124, the crack extension deflection is limited with the position of the grain boundary to $-75.186° < \Delta\theta_{124} < 104.814°$.

Table 1. Basic statistical parameters of the CTOD and CTSD values

Case	CTOD [μm]				CTSD [μm]			
	Min	Max	Mean	Std Dev	Min	Max	Mean	Std Dev
9.735, random	0.326	1.480	0.748	0.221	0.191	1.320	0.637	0.204
2R: 9.735, random	0.471	1.528	0.843	0.186	0.382	1.181	0.710	0.143
3R: 9.735, random	0.510	1.447	0.876	0.172	0.539	1.152	0.775	0.114
4R: 9.735, random	0.586	1.412	0.923	0.163	0.519	1.156	0.785	0.119
6R: 9.735, random	0.751	1.406	1.058	0.143	0.615	1.172	0.865	0.112
Monocrystal	0.116	2.131			0.097	1.838		
9.735, 135, random	0.304	1.407	0.735	0.223	0.181	1.039	0.501	0.162
9.735, 2R: 135, random	0.265	1.090	0.602	0.168	0.127	0.923	0.496	0.145
9.735, 3R: 135, random	0.203	0.883	0.530	0.144	0.133	0.645	0.367	0.118
9.735, 4R: 135, random	0.163	0.752	0.445	0.124	0.138	0.606	0.351	0.107
9.735, 6R: 135, random	0.139	0.456	0.289	0.073	0.076	0.410	0.220	0.071
Bicrystal, 9.735, 135	0.134				0.095			

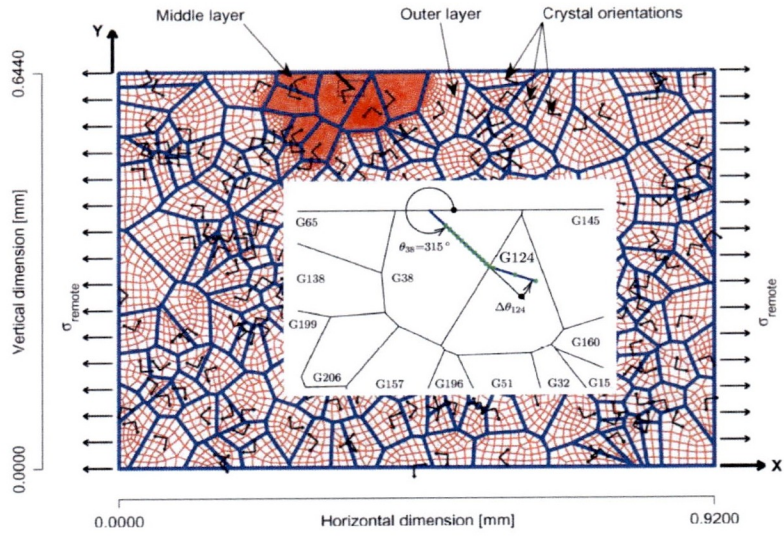

Figure 16. Setup of a set of stationary cracks mimicking the growth towards and across the grain boundary. (Reprinted from Simonovski and Cizelj [40] © permission from Elsevier.)

The orientation of the lattice in the subsequent grain should be consistent with the crack growth direction. The crack could be placed within two active slip planes, P2 or P4, see Figure 5. The corresponding lattice orientations are therefore defined as:

$$\alpha_{P2} = (315° + \Delta\theta_{124}) - 180° - 90° - 35.264° \text{ and} \qquad (39)$$

$$\alpha_{P4} = (315° + \Delta\theta_{124}) - 180° - 90° + 35.264°. \qquad (40)$$

By analogy with the stationary oblique crack, it is deemed practical to perform a sensitivity study of CTOD and CTSD for typical orientations within a bicrystal environment before embarking on the random polycrystal case.

Bicrystal

As already mentioned, the orientation of the crack in grain #38 is fixed at $\theta_{38}=135°$. The crack extension deflection and the corresponding lattice orientations in the grain #124 is then systematically varied from -75.186° to 104.814° in increments of approximately 2°. The maximum length of the kinked crack extension into grain #124 is 30.39 μm. The size of the grain #124, estimated as the square root of its area, is equal to 60.78 μm, which represents twice the length of the longest crack kink.

The CTOD as a function of the crack extension deflection is depicted in Figure 17. Please note that each crack extension deflection assumed two different lattice orientation accommodating crack within slip plane P2 or P4. These resulted in two CTOD curves: full for the crack in P2 and dashed for the crack in P4.

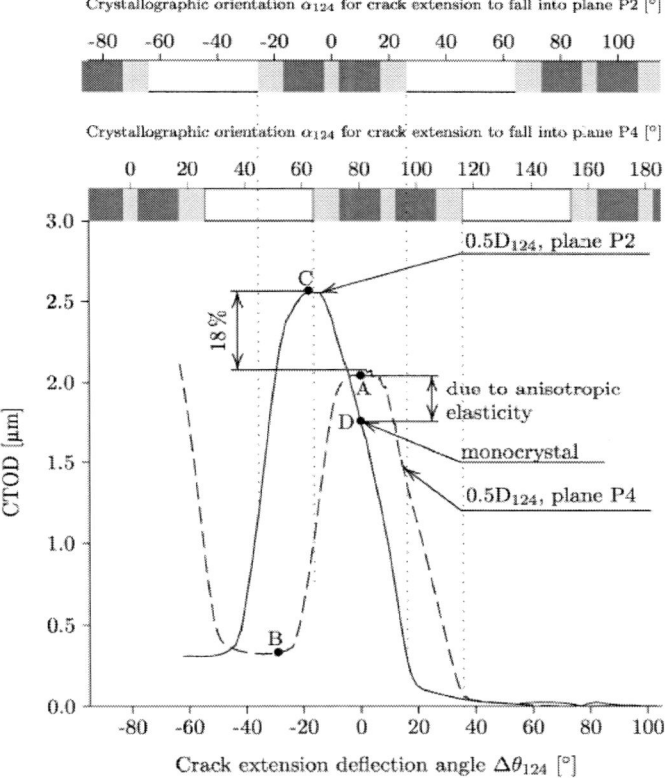

Figure 17. CTOD as function of crack extension deflection at remote load 240 MPa. (Reprinted from Simonovski *et al.*, [39] © permission from Elsevier.)

Two shaded bars on the top of the figure represent the grain hardness index, introduced in Figure 7. These two bars are shifted according to whether the crack extension falls into the slip plane P2 or P4. The corresponding crystallographic orientations are indicated next to the bars. These orientations are calculated using eqs. (39) and (40). Label $0.5D_{124}$ stands for the crack extension length being half the size of the grain 124.

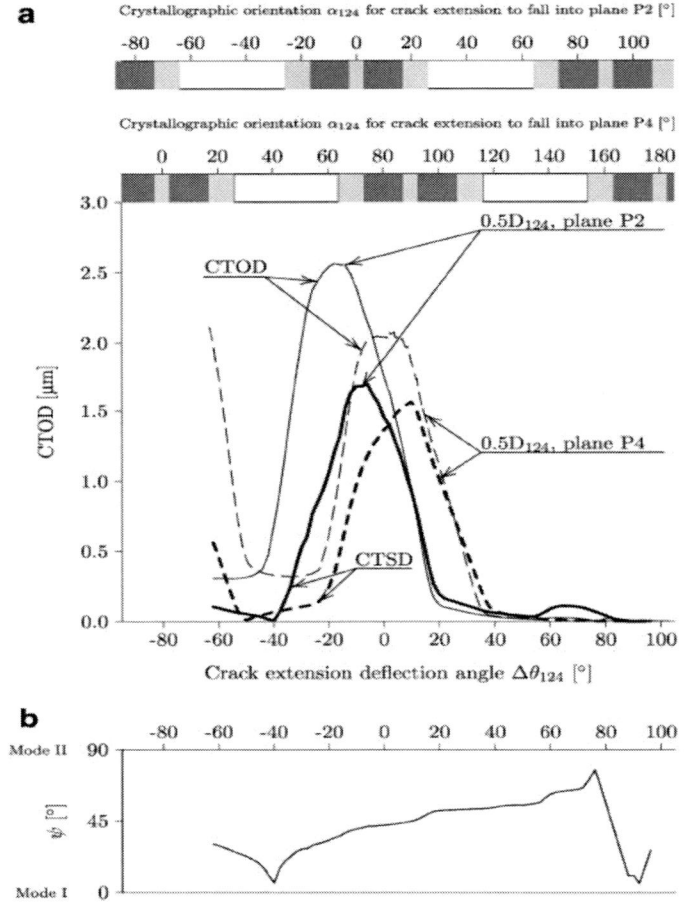

Figure 18. (a) CTOD and CTSD as function of crack extension deflection and (b) corresponding mode-mixity parameter ψ at remote load 240 MPa. (Reprinted from Simonovski et al., [39] © permission from Elsevier.)

One can immediately observe that the highest CTODs (points A and C) are obtained at crystallographic orientations with soft elasto-plastic response, whereas the small CTODs are obtained where the response is purely elastic. The lowest CTODs are obtained at the crack extension deflection $\Delta\theta_{124} = 38°$, where the crack extension comes close to being parallel to the remote external load. The low CTOD values are therefore expected in this region.

As the crack deflection increases above 45°, the CTOD values are additionally reduced due to a shift of the highest equivalent strains from the crack tip to the crack kinking point. Similar effect was noticed for crack extension placed in slip plane P4.

If the crack extension deflection is 0°, the crack does not kink. If additionally, the crack is placed in the slip plane P2 (point D), the crystallographic orientations are identical for both grains and we have a monocrystal configuration. Note that CTOD in this case ($\Delta\theta_{124}$ = 0°) is actually smaller than when the crack is placed in the P4 slip plane (dashed line). The reason for this is in the orientation of the crystal containing crack extension. To place the crack extension into the P4, the crystal is rotated at 80.264°, which results in a softer response compared to 9.736° (crack in P2, Figure 7). Since the largest Schmid factors are the same for both angles of rotation, the main contributing factor resulting in this CTOD difference is caused by the anisotropic elasticity.

A difference of 18% in maximal CTOD values, see Figure 17, is attributed to the difference in crack extension direction and slip plane. Maximal CTOD for crack extension in plane P4 is attained at crack extension deflection of -18° compared to 2° for the crack extension placed into the plane P2. The configuration with -18° is nearly perpendicular to external load and therefore naturally results in larger CTOD. For a reference, a difference of 24% in CTOD between the two crack extension directions was obtained using equivalent isotropic elastic material model.

CTSD values are depicted along the CTOD values in Figure 18 (a). It is evident that crystallographic orientations with soft elasto-plastic response also result in high CTSD values, while orientations resulting in pure elastic response decrease CTSD values. For the most of the configurations the CTOD and CTSD are only slightly shifted, which implies mixed-mode crack. The mode-mixity parameter ψ depicted in Figure 18 (b) is based on small scale yielding definition:

$$\Psi = \arctan\sqrt{\frac{CTSD}{CTOD}}. \qquad (41)$$

$\psi = 0°$ corresponds to pure Mode I, whereas $\psi = 90°$ corresponds to pure Mode II. The crack extensions deflections for which maximum CTOD are obtained are clearly in mixed mode.

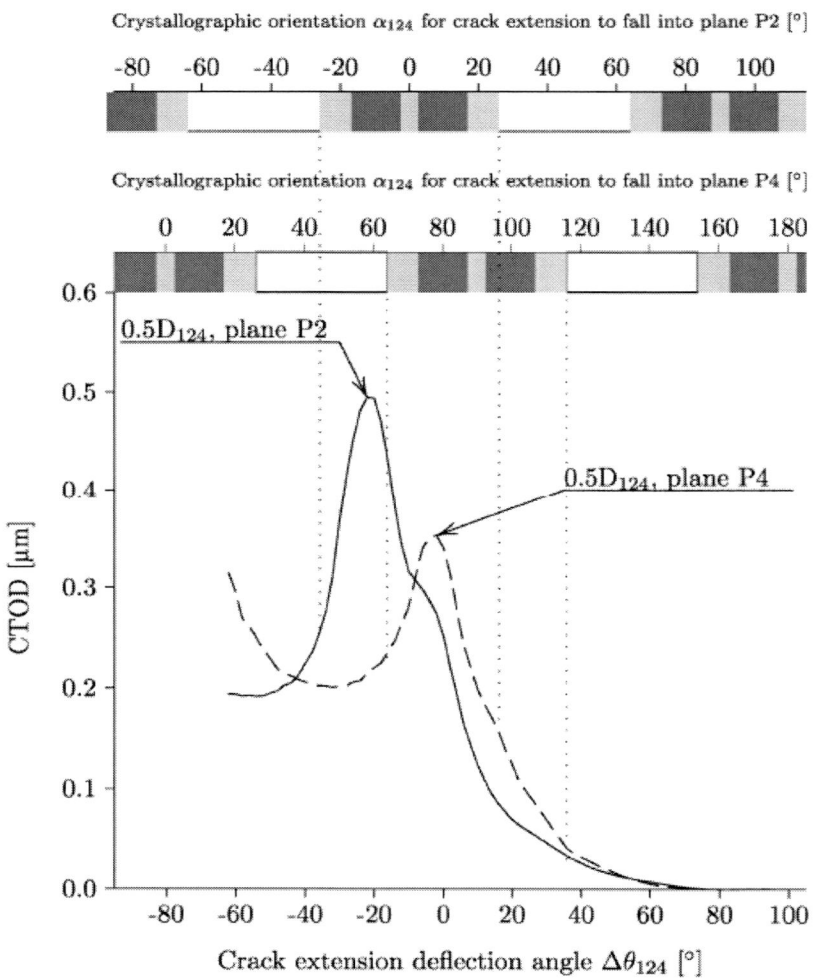

Figure 19. CTOD as function of crack extension deflection at remote load 192 MPa. (Reprinted from Simonovski et al., [39] © permission from Elsevier.)

The conclusions made so far are also valid for smaller external load of 192 MPa (Figure 19). One can see that a small reduction in load resulted in almost 4 times smaller CTOD values. Maximal CTODs are again obtained at crystallographic orientations resulting in soft elasto-plastic response. However, due to smaller amount of activated slip, areas of high CTODs are narrower compared to larger load, Figure 17. Also, the relative difference between the peaks is now larger due to the increased elastic contribution to the overall deformation. Analysis with shorter crack extension lengths was also performed and fully supports the conclusions [39].

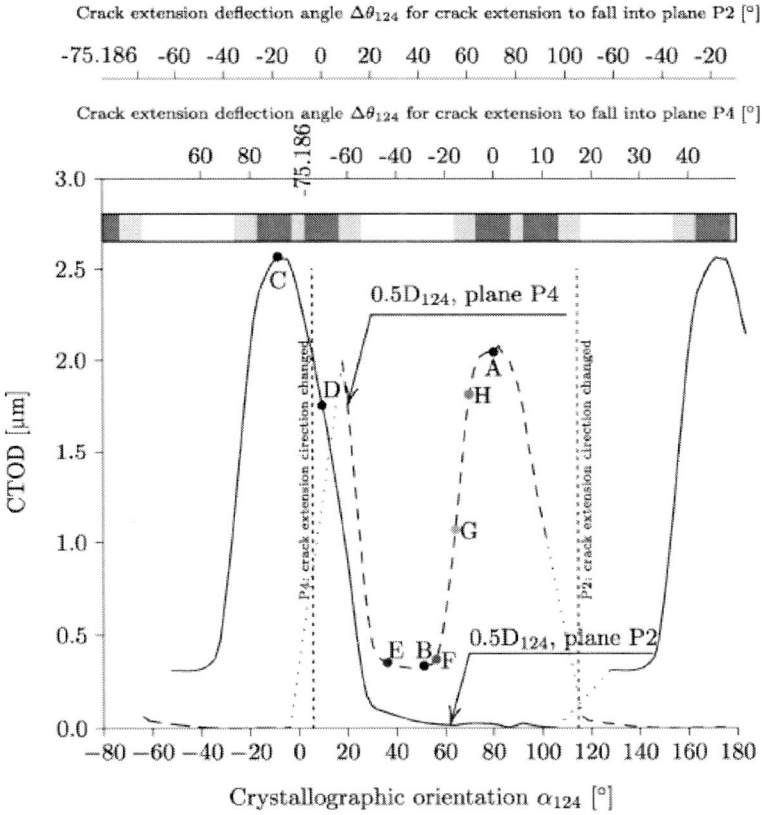

Figure 20. CTOD as function of crystallographic orientation of grain #124 at remote load 240 MPa. (Reprinted from Simonovski et al., [39] © permission from Elsevier.)

It might be more convenient to present the computed crack opening displacements as a function of the crystallographic orientation of the grain #124 rather than the crack extension deflection angle. For a given crystallographic orientation, the Stage I crack will align with one of the slip planes (P2 or P4). In Figure 20, the CTOD is plotted versus the crystallographic orientation of grain #124 for the same load and crack length as in Figure 17. The maximum value for P2 (P4) is attained at $\alpha_{124} = -8.264°$ ($\alpha_{124} = 82.264°$) with the associated crack extension deflection $\Delta\theta_{124} = -18°$ ($\Delta\theta_{124} = 2°$).

Dashed lines in Figure 20 indicate the orientations where crack direction coincides with the grain boundary. Meshes of crack orientations nearly parallel with grain boundary were not generated due to constraints in the shape of the quadrilateral elements used in the mesh. The missing CTOD values are indicated with dotted lines.

It is reasonable to assume that a crack crossing the grain boundary into a given crystallographic orientation will select the slip plane where the crack tip opening displacement is the highest. It is to be noted that the CTOD values due to different crystallographic orientation can differ up to a factor of 7.7 (compare points B and C). This further implies that the crack would propagate in plane P4 for 13.7° 6 < α_{124} < 114.550° and in P2 for other grain orientations. It can also be inferred that the crack extension deflection angle will always be -75.141° < $\Delta\theta_{124}$ < 34° [39].

The crack will generally be loaded in mixed mode. This requires that the criterion for selecting the slip plane in which the crack aligns should also be based on some complex mixed mode criterion. However, since the CTOD and CTSD are more or less in phase, a simple criterion based on CTOD might give qualitatively reasonable prediction of the crack extension deflection angles.

Polycrystal Configurations

Five polycrystal configurations have been analyzed. Crystallographic orientation for the first two grains containing the crack (grains 38 and 124) are taken from orientations corresponding to points A, E, F, G, and H of the bicrystal case in Figure 20. The relevant orientations of lattice and crack are summarized in Table 2 and visualized in Figure 21.

All remaining grains assume random crystallographic orientations with uniform distribution in the range 0 to 2π. All together 100 different random sets were generated for each of the five polycrystal configurations.

Table 2. Crystallographic orientations for the grain 124 and corresponding crack deflection angles

Crystallographic orientation		Crack deflection angle $\Delta\theta_{124}$		Point in Figure 20
α_{38}	α_{124}	Crack in P2	Crack in P4	
9.735°	36.264°	26.528°	-44.000°	E
9.735°	50.857°	41.121°	-29.407°	B
9.735°	56.264°	46.528°	-24.000°	F
9.735°	64.264°	54.528°	-16.000°	G
9.735°	70.264°	60.528°	-10.000°	H
9.735°	80.000°	70.264°	-0.264°	A

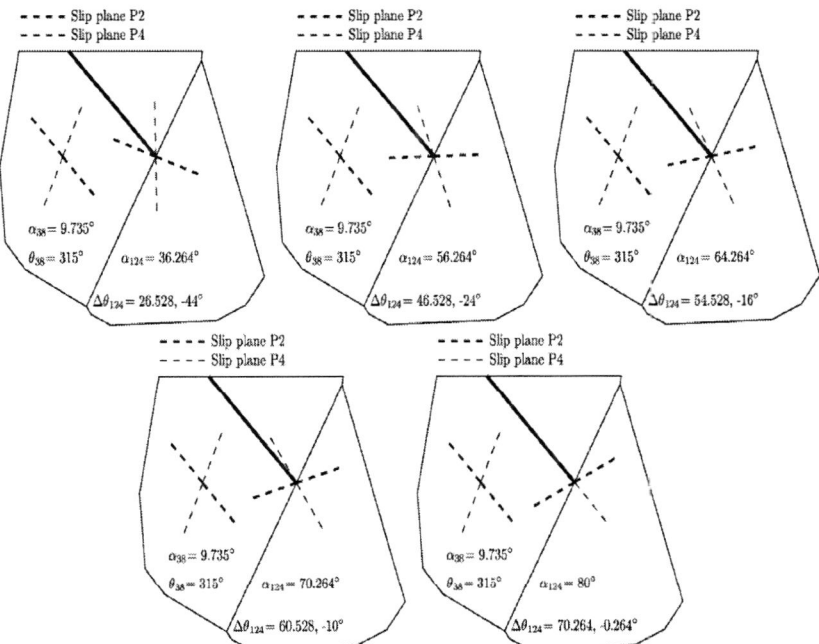

Figure 21. Selected orientations of the lattice beyond the first grain boundary at remote load 240 MPa. (Reprinted from Simonovski and Cizelj [40]T © permission from Elsevier.)

Crossing Grain Boundary

In this section, we examine the crack tip displacements as the crack is extended from the first into the second grain. For this purpose a series of models with embedded stationary cracks of different lengths was created (see Figure 16).

Crack length in the grain #38 varies from 17.72 μm to 52.37 μm (0.25 to 0.739 of the grain length of 70.87 μm). Once the crack extends across the grain boundary into the grain 124, its length in the grain #124 extends up to 30.39 μm (half of the grain length of 60.78 μm). Several crystallographic orientations of the grain #124 were used while placing the crack in either slip plane P2 or P4, see Table 2 and Figure 21.

Figure 22 shows the CTOD and CTSD displacements for crack in grain 124 placed in the slip plane P2. Values on abscissa indicate the length of the crack relative to the length of grain #38: value of 0 indicates the grain boundary. One can see that different crystallographic orientations change CTOD values of cracks approaching the grain boundary by up to 26% for the case with the shortest crack. As the crack is extended into the grain 124 this effect becomes much more pronounced. This is to be expected since the crack has to change its direction at the grain boundary. It was observed that when the crack direction turns upward to follow the slip plane P2, the maximal equivalent strain gradually shifts from the crack tip to the crack kink point. This results in up to 10 times smaller crack tip displacements as the crack crosses the grain boundary. Another important observation is that in all analyzed cases significant CTSD values were observed, confirming again the already noted mixed mode loading.

Figure 23 shows crack tip displacements when the crack in grain 124 is placed in the slip plane P4. In these cases the crack in grain 124 is almost perpendicular to the external load. In fact, we see that the closer the crack kink is to being perpendicular to the external load, larger the CTODs and lower the CTSDs are. This would suggest that among all the available slip planes the crack would probably propagate through the slip plane that is more perpendicular to the external load. This is also in line with well the known observation that short cracks gradually change from Stage I to Stage II where their direction is perpendicular to the external load.

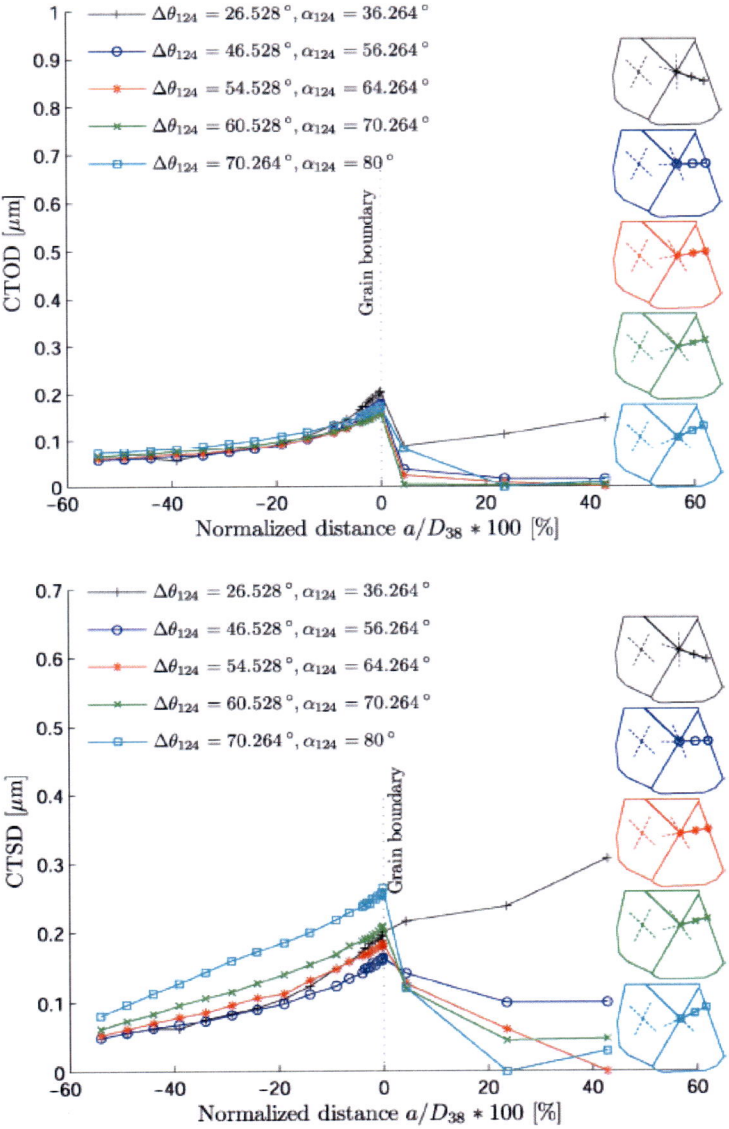

Figure 22. CTOD and CTSD of a growing crack as a function of crystallographic orientation (slip plane P2, remote load 240 MPa). (Reprinted from Simonovski and Cizelj [40] © permission from Elsevier.)

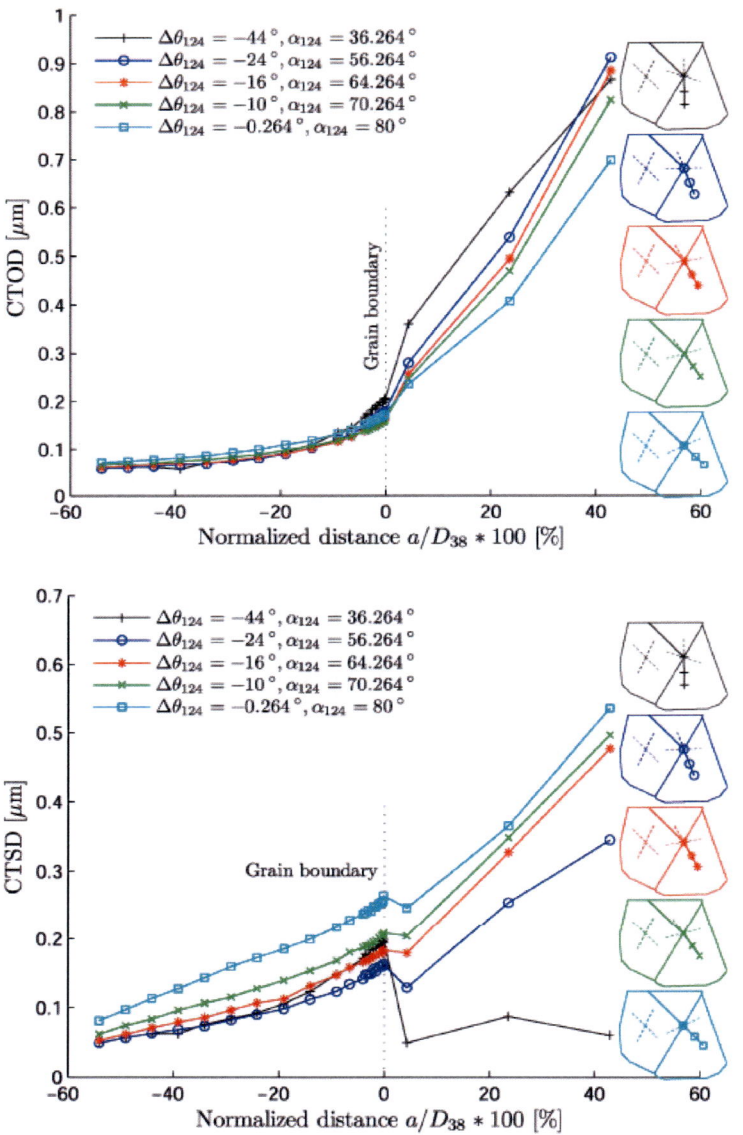

Figure 23. CTOD and CTSD of a growing crack as a function of crystallographic orientation (slip plane P4, remote load 240 MPa). (Reprinted from Simonovski and Cizelj [40] © permission from Elsevier.)

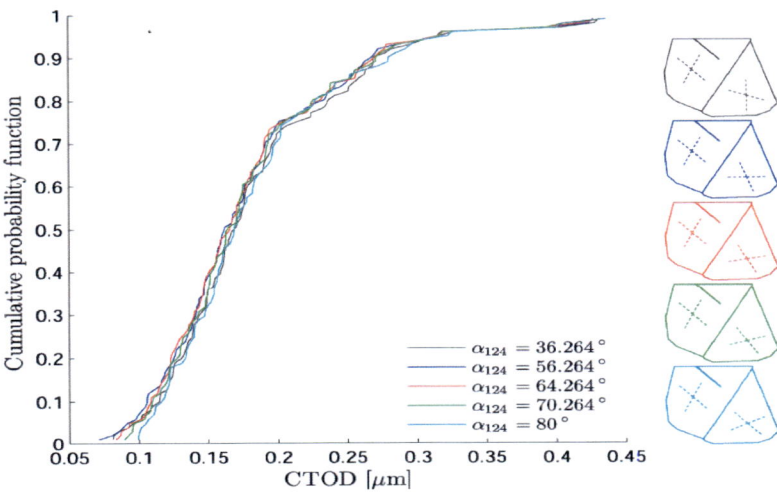

Figure 24. Cumulative probability of CTOD of oblique cracks (crack in slip plane P4, remote load 240 MPa). (Reprinted from Simonovski and Cizelj [40] © permission from Elsevier.)

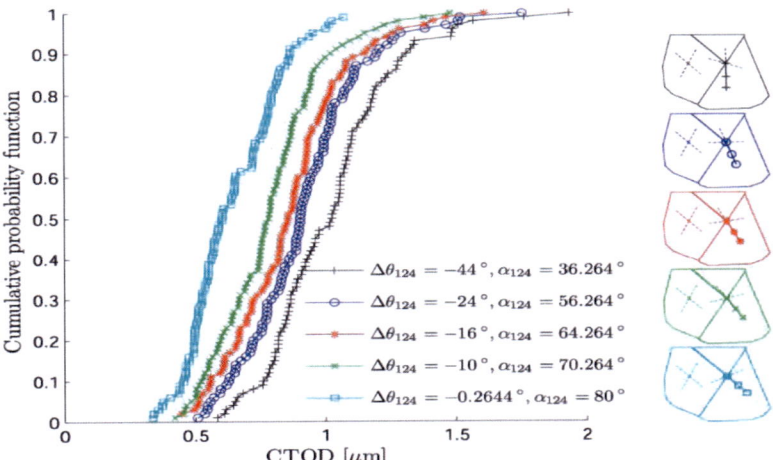

Figure 25. Cumulative probability of CTOD of kinked cracks (kink in slip plane P2, remote load 240 MPa). (Reprinted from Simonovski and Cizelj [40] © permission from Elsevier.)

The cumulative probability of CTOD for a oblique crack is depicted in Figure 24. These results are similar to those in Figure 12. Two important differences however exist. The first one is lower remote load of 240 MPa in

Figure 24 as compared to the 280 MPa in Figure 12. The second one is controlled orientation of lattice in grain #124: results in Figure 24 were obtained using the fixed orientations outlined in Table 2 and Figure 21, whereas fully random setting is presented as a part of Figure 12. Qualitative agreement between Figure 12 and Figure 24 is considered adequate.

Figure 25 depicts the cumulative probabilities of kinked cracks. The orientations of the lattice in grain #124 again followed those in Table 2 and Figure 21. The probability curves are rather close in Figure 24, indicating relatively small influence of the orientation of the grain #124 on the CTOD values. However, the scatter of the results along the abscissa shows that the orientations of grains beyond grains 38 and 124 have a significant impact. This is fully consistent with findings in Figure 12.

The impact of lattice orientation in grain #124 however increases significantly once the crack kink is extended into the grain #124. This can be seen as a larger distance between different lines in Figure 25. On the other hand, the impact of orientations of grains beyond #38 and #124 on the CTOD values decreases. This is manifested in decreased scatter of values along abscissa.

Two interlinked factors should also be mentioned once the crack is extended into the grain #124. The first is crack deflection. Larger CTODs are obtained when the crack extension in grain #124 is more perpendicular to the external load. Additionally, the crystallographic orientation also affects the stiffness of grain #124. $\Delta\theta_{124} = 36.264°$ results in the lowest Schmid factors among the analyzed configurations. The CTODs are however the highest in this configuration. This is attributed to the fact that the crack is the most perpendicular to the external load. Increasing the $\Delta\theta_{124}$ would cause the Schmid factors to increase and the crack extension to move away from being perpendicular to the external load. As a result, the values of the CTOD decrease. The crack extension direction in this case seems to be the main factor influencing the CTOD.

CRACK CROSSING MULTIPLE GRAIN BOUNDARIES

In this section the approach to estimate the crack length causing vanishing influence from the microstructural feature is described. This is achieved through a model containing 5027 randomly sized, shaped and oriented grains (Figure 26).

Computational Setup

A series of transgranular cracks of lengths from 1 to 7 typical grain lengths are inserted into the model, extending from a grain at the surface and kinking across grain boundaries towards the interior of the model. All of them are assumed to be stationary snapshots of a Stage I fatigue crack. The shortest crack is completely embedded in the grain No. 4854 (Figure 27). The crack is placed in the slip plane denoted as P2 in Figure 4. Increasing the crack length requires crossing the grain boundaries and kinking of the crack direction, which is assumed to follow the slip plane also in the newly damaged grain. In the model, the crack path was assumed arbitrarily, keeping in mind that the growing crack would finally tend towards the pure mode I loading (crack direction of 270°). The assumed path of the crack and the assumed coincidence of the crack with slip plane P2 (Figure 4) also required that the orientations of the lattice in grains along the crack path are fixed accordingly (see eq. (**Error! Reference source not found.**)).

Schematics of the crack path and related finite element mesh are given in Figure 27. Table 3 lists the crystallographic orientations α and crack directions θ for all grains along the assumed crack path. The orientation of all remaining grains in the model is random with uniform distribution in the range between 0 and 2π.

Equivalent Crack

The shape of the crack assessed in this section is rather complex. It is therefore useful to introduce a suitable effective crack shape with available linear elastic fracture mechanics (LEFM) solution. Two immediate purposes of the effective crack shape and length come in mind:

- Availability of reliable reference values of LEFM crack tip opening displacement (CTOD) and
- Proportionality of the CTOD to the crack length in LEFM.

Two crack shapes with available solutions [64] have been selected: kinked and oblique cracks emanating from a surface of a semi infinite plate, as depicted in Figure 28. It turns out that the difference in stress intensity factors between kinked and oblique cracks in similar configurations (see Figure 28) is of the order of 1% or less. Simpler oblique crack has been therefore selected as the representative effective crack. The appropriate lengths of effective cracks are listed in Table 3.

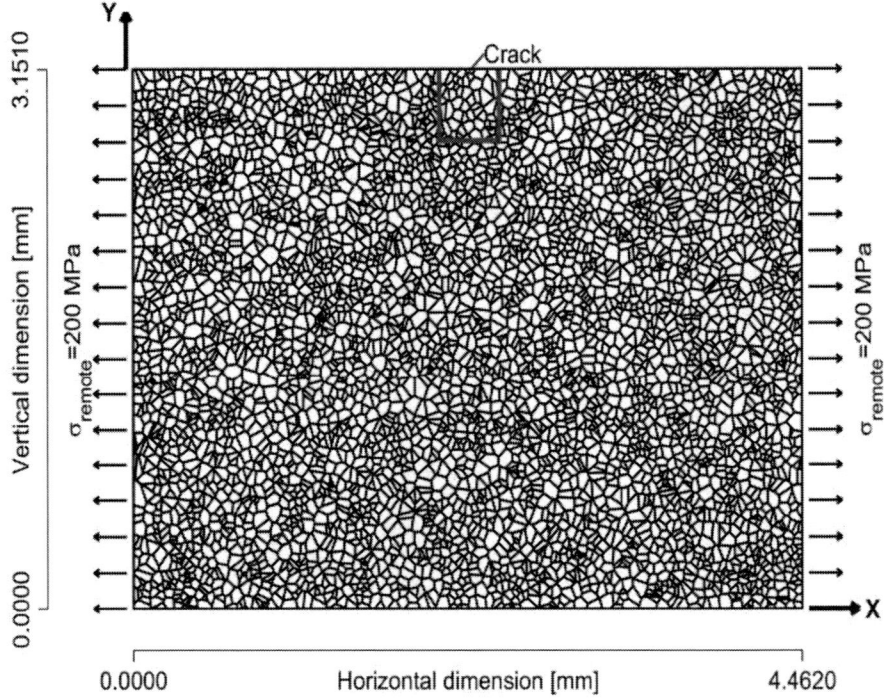

Figure 26. The outline of the finite element model with 5027 grains. (Reprinted from Cizelj and Simonovski [41] © permission from American Society of Mechanical Engineers.)

It should be noted here that the physical crack length is not always a good measure of the CTOD: crack ending in grain No. 88 is physically shorter than crack ending in grain No. 4956, but nevertheless results in much higher elastic CTOD.

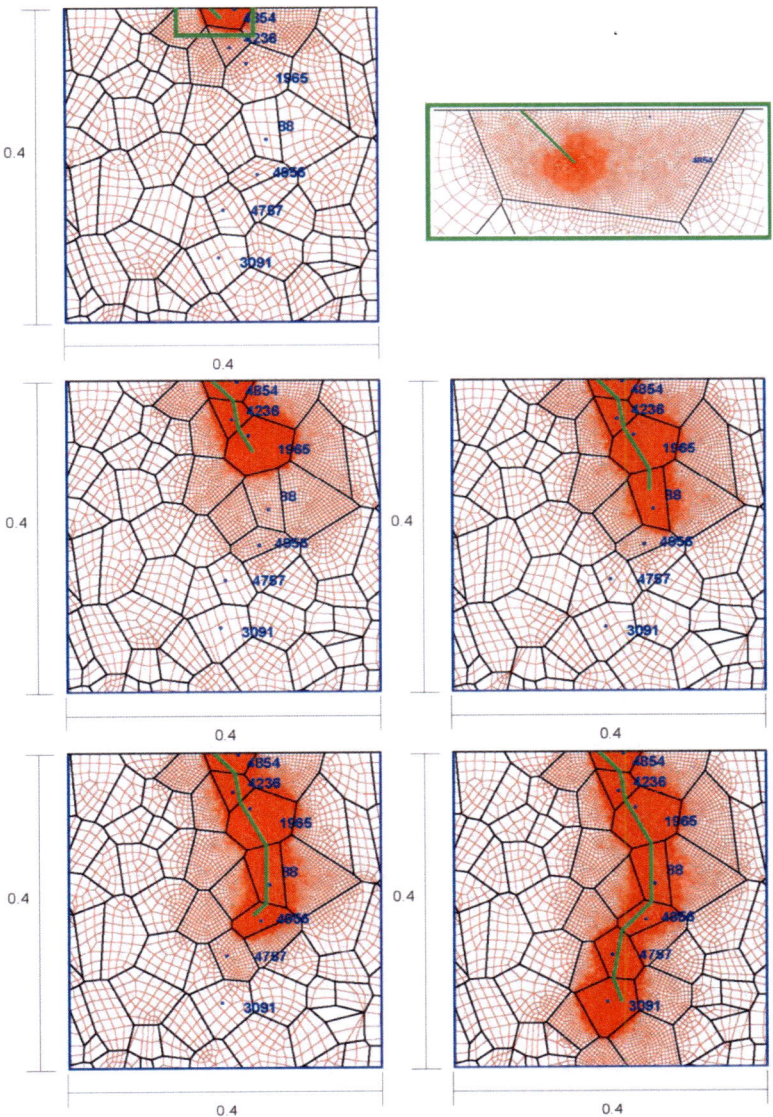

Figure 27. Details of the crack tip meshes for different crack lengths. Numbers indicate the grains containing the crack. Please refer to Figure 26 for the global position of the crack in the model. (Reprinted from Cizelj and Simonovski [41] © permission from American Society of Mechanical Engineers.)

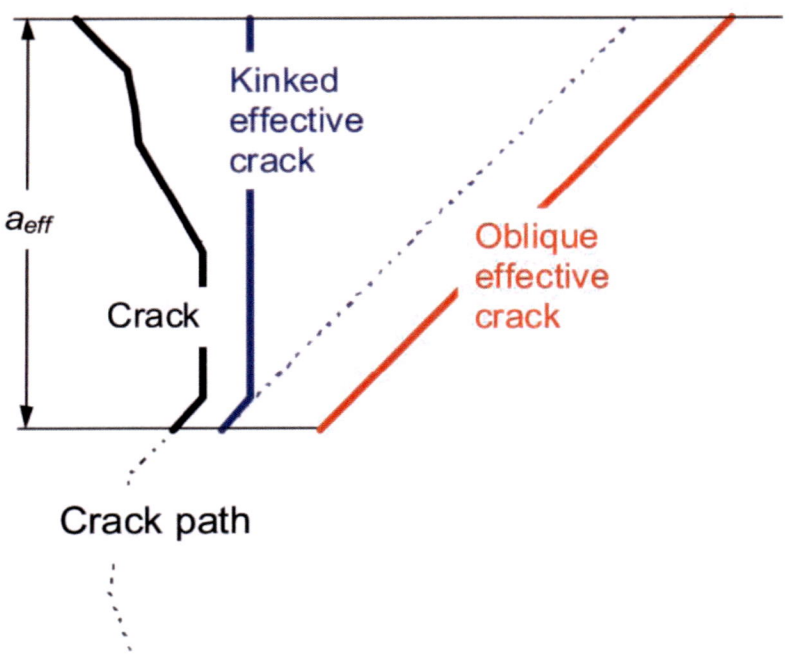

Figure 28. Possible effective crack shapes and effective crack length a_{eff}. (Reprinted from Cizelj and Simonovski [41] © permission from American Society of Mechanical Engineers.)

Table 3. Crack and crystal orientations and effective crack lengths

Grain No.	Crystalographic orientation α [°]	Crack direction θ [°]	Crack length [mm]			CTOD calculated
			Physical	Physical perpendicular to load	Effective a_{eff}	
4854	9.375	315	0.0187	0.0132	0.0093	Yes
4236	-25.264	280	0.0578	0.0465	0.0569	No
1965	-5.264	300	0.1114	0.0951	0.0929	Yes
88	-35.264	270	0.1693	0.1490	0.1874	Yes
4956	-80.264	225	0.2398	0.2127	0.1495	Yes
4757	-45.264	260	0.2974	0.2609	0.3193	No
3091	-15.264	290	0.3671	0.3281	0.3694	Yes

Results

A set of 30 models with different random grain orientations were generated for the five different crack lengths depicted in Figure 27 and detailed in Table 3. For each of the crack lengths, the average and standard deviation values of CTOD were then calculated. The 30 input values are sufficient to obtain statistically reasonably stable estimates of averages and standard deviations. It was expected that, with increasing crack length, the CTOD would depend less upon the surrounding microstructural features. Consequently, since the scatter of the CTOD is caused by the random orientations of the grains in the model, the value of the CTOD standard deviation is expected to decrease.

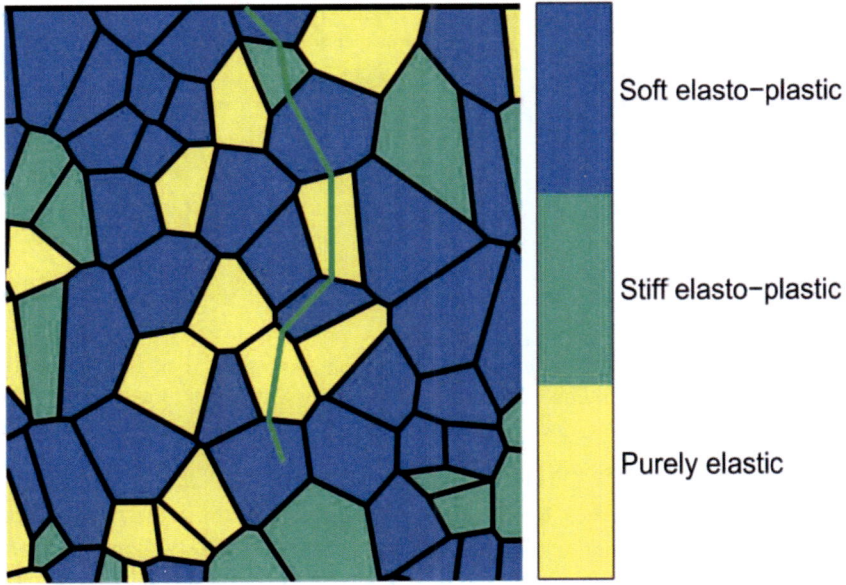

Figure 29. Grain stiffness for an arbitrarily selected set of crystallographic orientations. Orientation of crack containing grains is fixed. (Reprinted from Cizelj and Simonovski [41] © permission from American Society of Mechanical Engineers.)

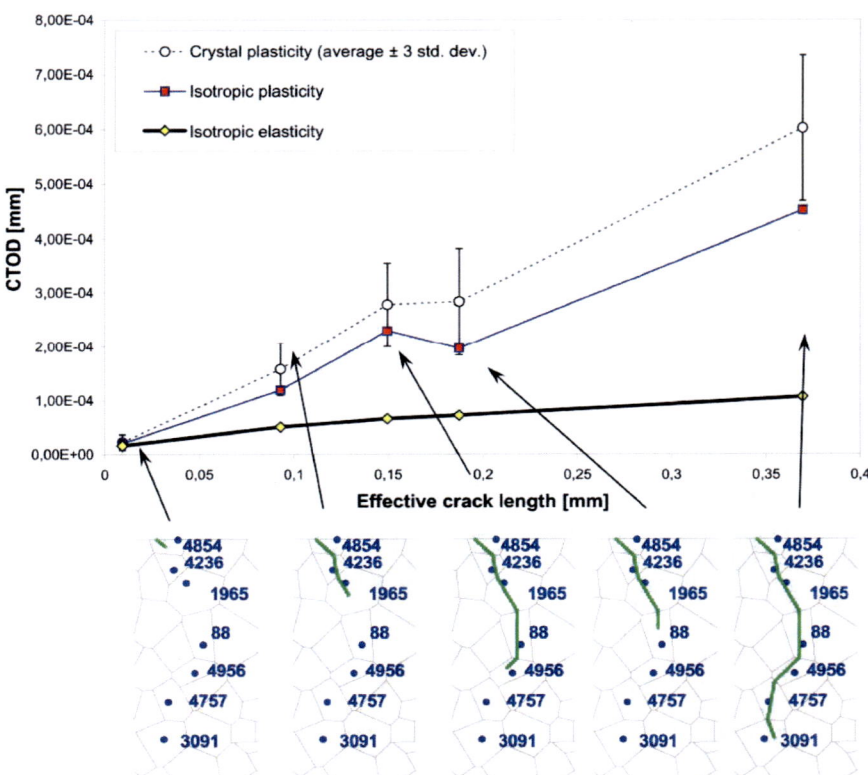

Figure 30. Scatter of the CTOD values due to the random nature of crystallographic orientations of the surrounding grains. (Reprinted from Cizelj and Simonovski [41] © permission from American Society of Mechanical Engineers.)

Figure 30 displays CTOD values as a function of the effective crack length. The effective crack length has been introduced to minimize the impact of the crack shape on the discussion of the results (see section Equivalent Crack). As expected, the CTOD values obtained using isotropic linear elastic material, are nearly proportional to the effective crack length. The small amount of non-proportionality is believed to be caused by rather complex crack shape, which could not be entirely captured by the effective crack approximation.

The mean values of the CTOD obtained using crystal plasticity and randomly oriented grains increase with the effective crack length. The exception is the stable CTOD value between the cracks with effective length of 0.1495 mm (crack tip in grain 4956) and 0.1874 mm (crack tip in grain 88).

Please note here that the crack with effective crack length of 0.15 mm is physically longer than the crack with effective crack length of 0.19 mm (Table 3). In addition, the relative stiffness of the grain containing the tip of the crack with effective length a_{eff}=0.19 mm is much higher than the relative stiffness of the grain containing the tip of the crack with a_{eff} =0.15 mm (compare Figure 8).

The scatter of the CTOD is indicated in Figure 9 with error bars denoting the range of ±3 standard deviations (~99% of all random realizations) and increases with the effective crack length.

The CTOD value obtained by the isotropic plasticity model is shown to represent an approximate lower limit. Both the isotropic elastic and isotropic plastic models are shown to significantly underestimate the CTOD values.

The normalized scatter of the CTOD is depicted in Figure 31 using the standard deviation as a measure of the scatter. Rather sharp decrease of scatter from above 20% to about 10% is noted for very short cracks. Further moderate decrease of scatter is seen with increasing the effective crack length. For the longest crack analyzed, which extends through seven grains, the standard deviation of the CTOD values is 7.3%.

The decrease of scatter was less pronounced than expected. Some possible reasons for this include:

- The crack was always placed in the P2 slip plane. In some cases, placing the crack in the P4 slip plane would result in the crack being more perpendicular to the external load, resulting in larger CTOD due to the remote load, which could mask some part of the scatter. This aspect was, however, not explored in this work. The emphasis was put on the evaluation of the scatter of the CTOD due to the crystallographic orientation.
- A micro-crack usually forms at surface extrusions/inclusions or along slip bands. The formation and position of these structures depends significantly on the crystallographic orientations of the grains. For different set of crystallographic orientations the position of these structures differs and the initial position of the crack should follow this. In our case the initial position of the crack is, however, arbitrarily assumed and fixed.
- The crack extending through 7 grains could be too short to become independent of the surrounding microstructural features. However, the longest crack in the model is longer than 10% of the model height.

Longer cracks would require model with larger number of grains, which is currently beyond our computational capabilities.

Based on the limited experience already gained with the 3D simulations [43], qualitatively similar results are expected for a 3D model, with possible quantitative change in the amount of scatter. Additional work is currently under way to verify this.

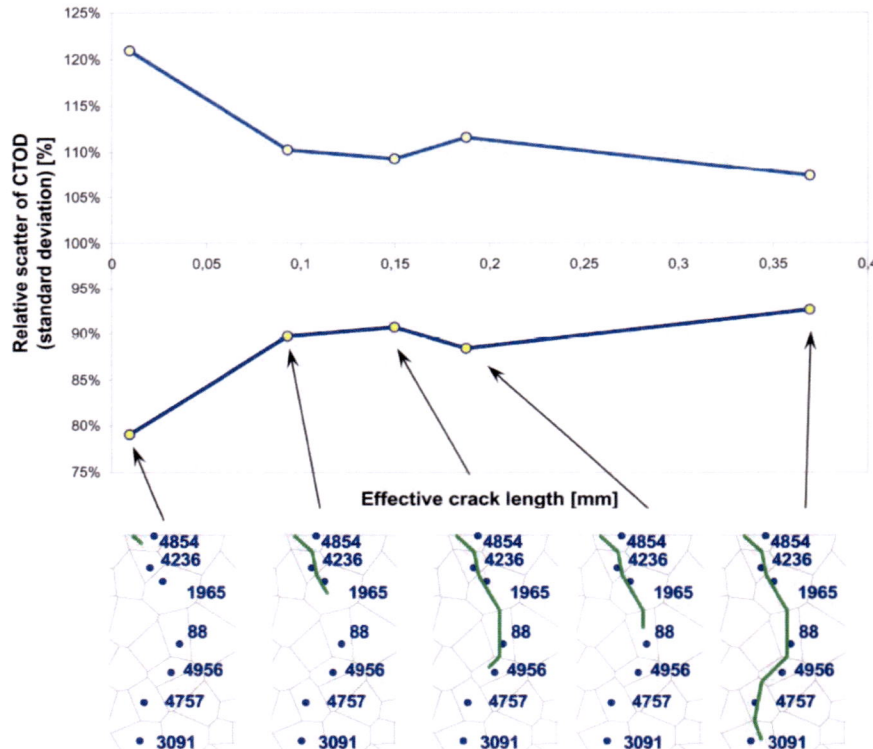

Figure 31. Relative scatter of the CTOD values due to the random nature of crystallographic orientations of the surrounding grains. (Reprinted from Cizelj and Simonovski [41] © permission from American Society of Mechanical Engineers.)

Chapter 4

OUTLOOK

This section summarizes some of the possible directions and challenges leading to possible improvements to be implemented in the future. Some preliminary results are also given. The selection has been made according to the potential to release the most stringent limitations of the computational setup proposed earlier.

PLANAR MODELS WITH SPATIAL MATERIAL ORIENTATION

Let us reconsider the short oblique crack discussed in section Stationary Oblique Surface Crack with fully spatial orientations of the lattice.

We will take the planar model shown in Figure 1 and extrude it to the thickness of 3.3 µm. The thickness is discretized with one finite element. The plane strain constraint is imposed by constraining the thickness of the model between two rigid planes. The finite elements used are 20 node reduced integration bricks (Figure 32).

In the planar model, the lattice orientation was fixed in the plane with random rotations permitted only around the normal to the plane of the model (Figure 4). This was done to ensure planar slip system activity (Figure 5). The setup in Figure 32 however, allows complete freedom in assigning the lattice orientations. This allows to verify the scatter of CTOD depicted in Figure 12. To achieve this, the orientation of the crack containing grain is left unchanged. The orientation of all other grains is assumed random and uniformly distributed within a solid angle of 4π.

Figure 32. Planar finite element model extruded to a thickness of 1 element.

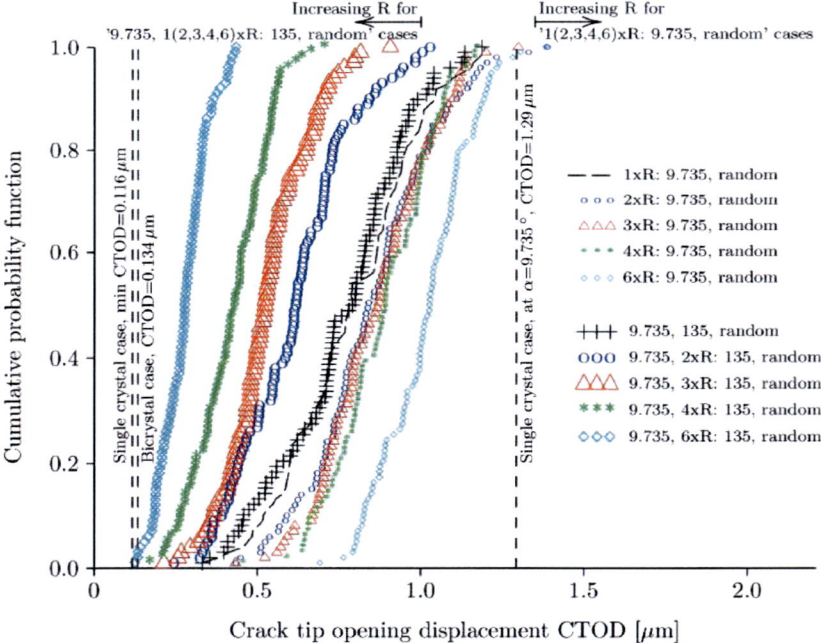

Figure 33. Cumulative functions of CTOD in planar and spatial random orientations compared.

The cumulative probability functions are formed from 100 simulations per case with random sets of lattice orientations and for CTOD depicted in Figure 33. The cumulative probabilities comply very well with those presented in Figure 12. It is noted however that the scatter obtained with planar lattice rotations (Figure 12) slightly over predicts the scatter obtained by spatial lattice orientations (Figure 33). Also, the effect of the slip bands is less pronounced with the spatial lattice rotations. Analogous results have been obtained for CTSD.

SPATIAL MODELS

Further improvement of planar models with spatial lattice orientation of lattice is of course an implementation of fully spatial models. Two possible families of models are briefly illustrated here.

Theoretical models such as for example spatial Voronoi tessellation may be used as approximation of grain shapes. This brings well defined borders between grains and rather straightforward book-keeping on the basic geometric model constituents (e.g., vortices, lines, etc). An example of spatial Voronoi tessellation meshed by tetrahedron finite elements, taken from [65], is depicted in Figure 34.

Figure 34. Spatial Voronoi tessellation with 524 grains, meshed with 437,009 thetrahedral finite elements.

"As measured" microstructures might represent another modeling approach. An example, is depicted in Figure 35. It is noted here that much more effort is needed to appropriately characterize the geometrical aspect of the model in this approach.

Figure 35. **Approximation of "as measured" microstructure**. Surfaces are approximated by 29928 triangles.

PROPAGATION OF CRACKS

Predicting the propagation of short cracks is certainly one of the most important future goals of the computational setup. A possible approach would be to model a special cohesive layer along the potential crack path, as illustrated schematically in Figure 36. Such approach would be especially promising in modeling intergranular cracks, where the potential crack path is completely defined by the layout of the grain boundaries. An example of damage model to be implemented in the cohesive layer is illustrated in Figure 37.

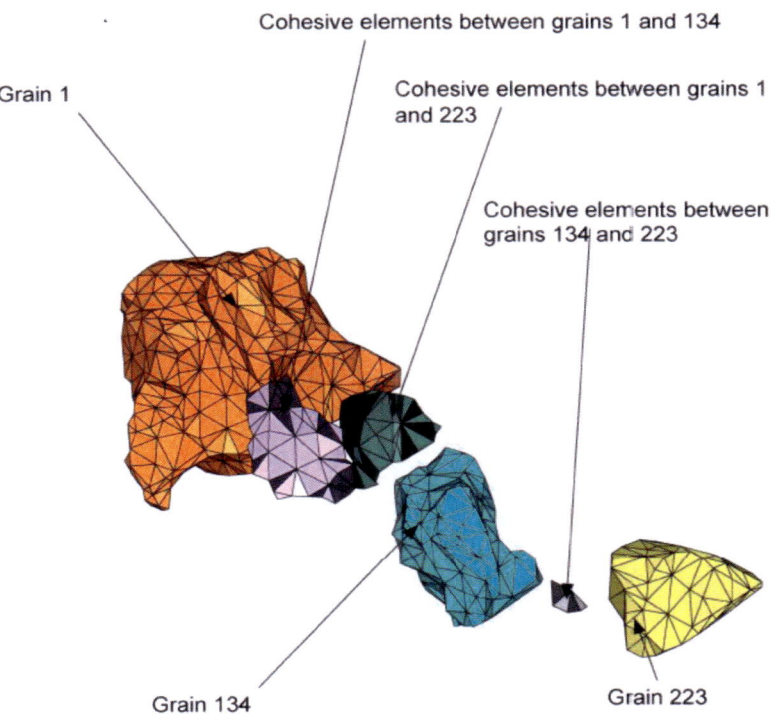

Figure 36. Principal layout of cohesive layers between grains, modeled using cohesive finite elements.

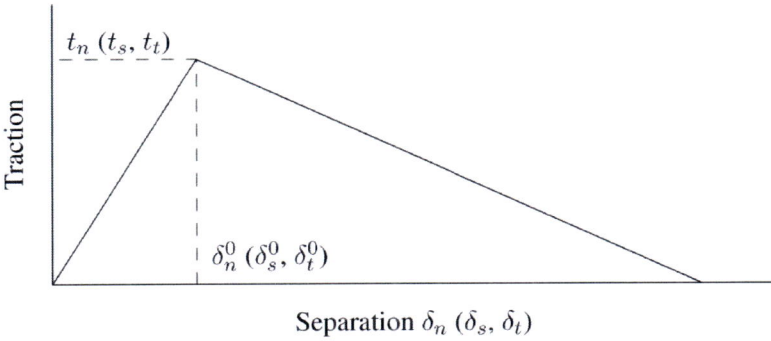

Figure 37. Basic description of damage behavior of a cohesive element using traction-separation measures.

Chapter 5

CONCLUSION

Computational model aiming to quantify the effects of random grain shapes and orientations on the variability of crack tip opening and sliding displacements (CTOD, CTSD) is proposed. The microstructurally short cracks with lengths up to about ten grains are considered, as they are known to be strongly influenced by the microstructural features in the vicinity of the crack tip.

The computational model relies on the random grain structure simulated using the Voronoi tessellation. The constitutive behavior of individual grains includes randomly oriented anisotropic elasticity and crystal plasticity. The equilibrium equations are solved with macroscopic boundary conditions at the scale of the component using commercially available finite element solver ABAQUS. The face centered cubic material with properties representing industry grade austenitic stainless steel is assumed with macroscopic uniaxial loading approaching macroscopic yield strength of the material.

Several configurations with stationary crack configurations were studied. They include transgranular crack extending through about half of a crystal grain, a series of cracks of different lengths simulating the short crack approaching and crossing the first grain boundary and a series of cracks with different lengths extending from one to over a few grains. The cracks are always assumed to be aligned within a slip plane. Sufficiently many simulations with different random grain orientations have been performed to estimate the average, variance and cumulative probability functions of CTOD and CTSD.

Extensive sensitivity studies have been performed studying the cracks embedded in large mono- and bicrystals with variable lattice orientations. The resulting impact of lattice orientations on the CTOD/CTSD values was applied to define the most favorable and unfavorable lattice orientations. Additional useful results include possible limits of the CTOD/CTSD variability.

Two main sources of the CTOD/CTSD variability for a short crack embedded in a random polycrystal were identified and discussed: (1) the random grain structure; and, (2) the strain localizations (e.g., slip bands) extending over the entire computational domain. It is shown that the strain localizations could be at least partially responsible for about 5-10% of the highest CTOD/CTSD values.

The changes in CTOD/CTSD while the crack is approaching and crossing the grain boundary are also systematically quantified as a function of the lattice orientations on both sides of the grain boundary. Both the lattice mismatch at the grain boundary and the relative direction of the macroscopic load are shown to have significant influence.

An attempt has been also made to quantify the decreasing influence of the random grain structure with increasing crack length. The standard deviation of the CTOD estimate is shown to fall below 10% when the crack length grows to about 7 grains. From the engineering point of view, the cracks in AISI 316L longer than about 10 grains could be safely analyzed with classical fracture mechanics methods.

Current limitation of the computational setup is a plane strain model with lattice rotations around the out of plane axis only. This limits the slip system activity to two active (inplane) slip systems. The main reason for such limitation is the computational intensity of the simulations. A limited amount of simulations with spatial lattice 3D material orientations has also been performed and confirm the plausibility of results obtained by planar simulations.

An outlook towards possible future improvements of the computational setup is given including modeling of as-measured and simulated spatial microstructures and propagation of intergranular cracks.

ACKNOWLEDGMENTS

Financial support of the Slovene Research Agency (www.arrs.si) through grants P2-0026, J2-9168, 1000-07-380042 (co sponsored by Commissariat à l'Énergie Atomique, France, www.cea.fr), "Modelling intragranular damage in polycrystalline materias" (co-sponsored by British Council) and V2-0375 (co-sponsored by the Slovene Nuclear Safety Administration www.ursjv.si) is gratefully acknowledged.

REFERENCES

[1] S. Pearson (1975). Initiation of fatigue cracks in commercial aluminum alloys and the subsequent propagation of very short cracks. *Eng. Fract. Mech. 7*, 235-247.

[2] H. Kitagawa and S. Takahashi (1976). Applicability of fracture mechanics to very small cracks or the cracks in the early stage. Proceedings of Second International Conference on Mechanical Behavior of Materials, ASM, Boston, pp. 627-631.

[3] K.J. Miller (1987). The behavior of short fatigue cracks and their initiation. Part II-A general summary, *Fatigue Fract. Eng. Mater. Struct. 10* (2), 93-113.

[4] K. Hussain (1997). Short fatigue crack behavior and analytical models: A review, *Eng. Fract. Mech. 58*(4), 327-354.

[5] K. Hussain, E., de los Rios, and A. Navarro (1993). A two-stage micromechanics model for short fatigue cracks. *Eng. Fract. Mech., 44*(3), 425-436.

[6] S. Suresh, *Fatigue of Materials*, Cambridge University Press (1991).

[7] H. Andersson and C. Persson, (2004). *In situ* SEM study of fatigue crack growth behavior in IN718, *Int. J. Fatigue. 26* (3), 211-219.

[8] P. Hansson and S. Melin, (2005).Dislocation-based modeling of the growth of a microstructurally short crack by single shear due to fatigue loading, *Intl. J. Fatigue. 27* (4), 347-356.

[9] O. Düber, B. Künkler, U. Krupp, H.-J. Christ, and C.-P. Fritzen, (2006). Experimental characterization and two-dimensional simulation of short-crack propagation in an austenitic-ferritic duplex steel. *Intl. J. Fatigue, 28*(9), 983-992.

[10] W. L. Morris, O. Buck, and H. L. Marcus, (1976). Fatigue crack initiation and early propagation in Al 2219-T851. *Metall. Trans. A. 7A*, 1161-1165.
[11] C. Blochwitz, W. Tirschler, and A.Weidner, (1982) The growth of small fatigue cracks in 7075-T6 aluminium alloy. *Fatigue Eng. Mater. Struct. 5*, 233-248.
[12] W. L. Morris (1980). The noncontinuum crack tip deformation behavior of surface microcracks. *Metall. Trans. A. 11A*, 1117-1123.
[13] V. Tvergaard, Y. Wei, and J.W. Hutchinson, (2001). Edge cracks in plastically deforming surface grains, *Eur. J. Mech. A–Solid 20*(5) , 731-738.
[14] T. Zhai, A. J. Wilkinson, and J. W. Martin, (2000). A crystallographic mechanism for fatigue crack propagation through grain boundaries. *Acta Materialia, 48*(20), 4917-4927.
[15] A. Vašek, J. Polák, and L. Obrtlík, (1996). Fatigue damage in two-step loading of 316L steel. II. Short crack growth, *Fatigue Fract. Eng. Mater. Struct. 19* (2–3),157-163.
[16] J. Polák, K. Obrtlík, and A. Vašek, (1997). Short crack growth kinetics and fatigue life of materials. *Mater. Sci. Eng.* A, pp. 234-236; 970-973.
[17] M. El. Haddad, K. Smith, and T. Topper, (1979). Fatigue crack propagation of short cracks. *J. Eng. Mater. Techn.-Trans. ASME 101*, 42-46.
[18] K. Tanaka, Y. Akiniwa, Y. Nakai, and R. Wei, (1986). Modeling of small fatigue crack growth interacting with grain boundary. *Eng. Fract. Mech. 24*, 803-819.
[19] A. Navarro and E. R. de los Rios (1988). A microstructurally-short fatigue crack growth equation. *Fatigue Fract. Eng. Mater. Struct., 11*, 383-396.
[20] K. S. Chan and J. Lankford (1983). A crack-tip strain model for the growth of small fatigue cracks. *Scripta Metall., 17*, 529-532.
[21] P. D. Hobson (1982). The formulation of a crack growth equation for short cracks. *Fatigue Eng. Mater. Struct. 5*, 323-327.
[22] J. R. Rice, D. E. Hawk, and J. R. Asaro (1990). Crack tip fields in ductile crystals. *Intl. J. Fract. 42*, 301-321.
[23] J. Kysar (2000). Directional dependence of fracture in copper sapphire bicrystal. *Acta Mater. 48*, 3509-3524.
[24] K. Gall, H. Sehitoglu, and Y. Kadioglu (1996).Plastic zones and fatigue-crack closure under plane-strain double slip, *Metall. Mater. Trans. A 27A* , 3491-3501.

[25] K. Gall, H. Sehitoglu, and Y. Kadioglu (1996). FEM study of fatigue crack closure under double slip, *Acta Mater. 44* (10), 3955-3965.
[26] G. P. Potirniche, S. R. Daniewicz, and J.C. Newman, Jr. (2004). Simulating small crack growth behavior using crystal plasticity theory and finite element analysis, *Fatigue Fract. Eng. Mater. Struct. 27*(1), 59-71.
[27] V. P. Bennett and D. L. McDowell (2003). Crack tip displacements of microstructurally small surface cracks in single phase ductile polycrystals. *Eng. Fract. Mech. 70*, 185-207.
[28] G.P. Potirniche and S.R. Daniewicz (2003). Analysis of crack tip plasticity for microstructurally small cracks using crystal plasticity theory, *Eng. Fract. Mech. 70* (13), 1623-1643.
[29] R. Ballarini, R. L. Mullen, and A. H. Heuer (1999). The effects of heterogenity and anisotropy on the size effect in cracked polycrystalline films. *Intl. J. Fract. 95*, 19-39.
[30] Y. Wang and R. Ballarini (2002). A long crack penetrating a circular inhomogenity. *Meccanica. 38*, 579-593.
[31] M. Kovač (2004). Influence of microstructure of development of large deformations in reactor pressure vessel steel. Dissertation, Ph. D. Thesis, University of Ljubljana, Slovenia.
[32] M. Kovač and L. Cizelj (2005). Modeling elasto-plastic behavior of polycrystalline grain structure of steels at mesoscopic level. *Nucl. Eng. Des. 235*, 1939-1950.
[33] I. Simonovski, M. Kovač, and L. Cizelj (2004). Estimating the correlation length of inhomogeneities in a polycrystalline material. *Mater. Sci. Eng.* A, *381*, 273-280.
[34] T. Watanabe (1984). Approach to grain boundary design for strong and ductile polycrystals. ResMechanica: *Intl. J. Struct. Mech. Mater. Sci., 11*(1), 47-84.
[35] E. M. Lehockey, A. M. Brennenstuhl, and I. Thompson, (2004). On the relationship between grain boundary connectivity, coincident site lattice boundaries, and intergranular stress corrosion cracking". *Corr. Sci., 46*(10), 2383-2404.
[36] L. Cizelj and I. Kovše (2001). Short intergranular cracks in the piecewise anisotropic continuum model of the microstructure. Proceedings of the International Conference Nuclear Energy in Central Europe. Ljubljana: Nuclear Society of Slovenia.
[37] L. Cizelj and H. Riesch-Oppermann (2002). Towards growth model for short intergranular cracks in elastoplastic polycrystalline aggregate.

Proceedings of the International Symposium Contribution of Materials Investigation to the Resolution of Problems Encountered in Pressurized Water Reactors Fontevraud 5, (1),196-203, French Nuclear Society.

[38] I. Simonovski, K.-F. Nilsson, and L. Cizelj (2007). Crack tip displacements of microstructurally small cracks in 316L steel and their dependence on crystallographic orientations of grains. *Fatigue Fract. Eng. Mater. Struct., 30* , 463-478.

[39] I. Simonovski, K.-F. Nilsson, and L. Cizelj (2007). The influence of crystallographic orientation on crack tip displacements of microstructurally small, kinked crack crossing the grain boundary. *Comput. Mater. Sci, 39* , 817-828.

[40] I. Simonovski and L. Cizelj (2007). The influence of grains' crystallographic orientations on advancing short crack. *J. Fatigue 29* , 2005-2014.

[41] I. Simonovski and L. Cizelj, The influence of the grain structure size on microstructurally short cracks, *ASME J. Eng. Gas Turbines and Power* [in print].

[42] G. Kalinin, V. Barabash, S. Fabritsiev, H. Kawamura, I. Mazul, M. Ulrickson, C. Wu, and S. Zinkle (2001). ITER R&D: Vacuum vessel and in-vessel components: Materials development and test. *Fusion Eng. Design, 55*(2-3), 231-246.

[43] L. Cizelj and I. Simonovski (2008). Simulated planar polycrystals with planar and spatial random lattice orientations. Proceedings of the 9th Biennal ASME Engineering Systems Design and Analysis Conference, July 7-9, 2008, Haifa, Israel. The American Society of Mechanical Engineers.

[44] I. Simonovski and L. Cizelj (2008). Small cracks in a simulated columnar polycrystalline aggregate with random 2D and 3D lattice orientations, Proceedings of 1st International Conference on Engineering Against Fracture (ICEAF), Patras, Greece, May 28-30, 2008.

[45] Franz Aurenhammer(1991). Voronoi diagrams-A survey of a fundamental geometric data structure. *ACM Computing Surveys. 23*(3),345-405.

[46] Dietrich Stoyan, W.S. Kendall, and J. Mecke(1995). *Stochastic Geometry and Its Applications.* Chichester, England: John Wiley & Sons.

[47] H. Riesch-Oppermann (1999). Generation of 2D random Poisson–Voronoi mosaics as framework for micromechanical modeling of

polycrystalline materials-algorithm and subroutines description. Technical Report, FZKA 6325, Forschungszentrum Karlsruhe.
[48] S. Weyer, A. Fröhlich, H. Riesch-Oppermann, □. Heinz, L. Cizelj, M. Kovač (2002). Automatic finite element meshing of planar Voronoi tessellations. *Eng. Fract. Mech. 69* , 945-958.
[49] Stefan Weyer, (2001). Experimentelle Untersuchung und mikromechanische Modellierung des Schädigungsverhaltens von Aluminiumoxid unter Druckbeanspruchung. Dissertation. Karlsruhe, Germany: Universität Karlsruhe.
[50] J. F. Nye(1985). *Physical Properties of Crystals*. Oxford: Clarendon Press.
[51] G. I. Taylor (1938). Plastic strain in metals. *J. Instit. Metals, 62*, 307-324.
[52] J.R. Rice (1970). On the structure of stress–strain relations of time-dependent plastic deformation in metals, *J. App. Mech. 37* , 728-737.
[53] R. Hill and J.R. Rice (1972). Constitutive analysis of elastic–plastic crystals at arbitrary strain, *J. Mech. Phys. Solids. 20* (6) , 401-413.
[54] R. J. Asaro and J. R. Rice (1977). Strain localization in ductile single crystals. *J. Mech. Phys Solids. 25*(5),309-338.
[55] U. F. Kocks(1970). The relation between polycrystal deformation and single-crystal deformation. *Metall. Trans., 1,*1121-1143.
[56] R.J. Asaro (1983). Crystal plasticity, *J. App. Mech. 50* , 921-934.
[57] Y. Huang (1991). A user-material subroutine incorporating single crystal plasticity in the ABAQUS finite element program. Technical Report, Harvard University.(accessible through *http://www.columbia.edu/ ~jk2079/fem/umat_documentation.pdf*).
[58] D. Peirce, R. J. Asaro, and A. Needleman (1983). Material rate dependence and localized deformation in crystalline solids, *Acta Metall. 31* (12) , 1951-1976.
[59] T. Y. Wu,J. L. Bassani, and C. Laird (1991). Latent hardening in single crystals, I. Theory and experiments. *Proceedings of the Royal Society of London*. Series A. *435*,1-19.
[60] J. L. Bassani and T. Y. Wu (1991). Latent hardening in single crystals, II. Analytical characterization and prediction. *Proceedings of the Royal Society of London*. Series A. *435,*21-41.
[61] Sia. Nemat-Nasser(1999). Averaging theorems in finite deformation plasticity. *Mech. Mater. 31,*493-523.

[62] I. Simonovski, K.-F. Nilsson, and L. Cizelj (2005). Material properties calibration for 316L steel using polycrystalline model. International 13th Conference on Nuclear Engineering, May 16-20, 2005, Beijing, China.
[63] Igor Simonovski (2004). Mechanisms for thermal fatigue initiation and crack propagation in NPP components, Report No. 8948, "Jožef Stefan" Institute, Ljubljana, Slovenia.
[64] Y. Murakami, (1987). *The Stress Intensity Factor Handbook*, Pergamon Press.
[65] I. Simonovski, J. Marrow, L. Cizelj, J. Quinta de Fonseca, N. Stevens, and Y. Zhang (2008). Modeling of intergranular damage in polycrystalline metals, Proceeding of International Conference Nuclear Energy for New Europe, Portorož, Slovenia, Sept 8-11, 2008.

INDEX

A

acceleration, 4
ACM, 78
activation, 29
aging, 3
aggregates, 6
algorithm, 8, 78
alloys, 75
aluminium, 76
aluminum, 75
amplitude, 4
analytical models, 75
anisotropy, 77
arrest, 4
assignment, 8
averaging, 18

B

behavior, xi, 3, 4, 30, 70, 71, 76, 77
Beijing, 79
bias, 8
Boston, 75
boundary conditions, xi, 9, 71
boundary value problem, 5, 7
bounds, 35

C

calibration, 79
cell, 7, 8
Cellulose, v
China, 79
classical, 72
classification, 27
closure, 4, 76, 77
clusters, 31
compilation, 5
compliance, 1, 10
components, 6, 10, 14, 78, 80
configuration, 27, 37, 42, 55
Congress, viii
connectivity, 77
constraints, 45
convex, 8
cooling, 6
copper, 76
correlation, 5, 29, 77
corrosion, 77
crack, xi, xii, 2, 3, 4, 5, 7, 9, 20, 21, 22, 23, 25, 28, 29, 30, 31, 32, 33, 35, 37, 39, 40, 41, 42, 43, 44, 45, 46, 47, 49, 50, 51, 52, 54, 55, 56, 57, 58, 59, 60, 61, 62, 63, 65, 68, 71, 72, 75, 76, 77, 78, 80
cracking, 77

creep, 16
crystal lattice, 8, 9, 14
crystalline, 12, 17, 79
crystalline solids, 79
crystals, 5, 10, 20, 76, 79

D

definition, 22, 42
deformation, xi, 1, 2, 11, 12, 13, 29, 44, 76, 79
deviation, 60, 62
diffusion, 11
discretization, 8
displacement, 2, 5, 7, 23, 29, 45, 56
distribution, 31, 46, 56

E

elastic constants, 19
elastic fracture, xii, 2, 4, 56
elasticity, xi, 7, 10, 14, 26, 42, 71
energy, 77, 80
England, 78
environment, 39
equilibrium, xi, 71
Europe, 77, 80

F

fatigue, 3, 56, 75, 76, 77, 80
FCC, xii
February, 75
FEM, 77
films, 77
finite element method, 7
flow, 17
fracture, xii, 2, 4, 56, 72, 75, 76
France, 73
freedom, 65

G

generalization, 5
Germany, 79
goals, 68
grain, xi, xii, 1, 3, 4, 5, 7, 8, 9, 11, 19, 20, 22, 25, 26, 27, 30, 31, 32, 33, 35, 37, 39, 40, 41, 44, 45, 46, 47, 53, 54, 55, 56, 57, 58, 59, 60, 61, 62, 64, 66, 67, 68, 71, 72, 76, 77, 78
grain boundaries, 3, 5, 20, 56, 68, 76
grains, xi, xii, 3, 4, 5, 8, 20, 21, 22, 23, 25, 27, 28, 30, 31, 32, 33, 34, 35, 37, 39, 42, 45, 46, 47, 49, 50, 51, 54, 55, 56, 57, 58, 60, 61, 62, 63, 66, 67, 68, 69, 71, 72, 76, 78
grants, 73
Greece, 78
growth, 3, 5, 35, 37, 39, 75, 76, 77
growth rate, 3

H

Haifa, 78
hardening, 2, 16, 17, 19, 79
hardness, 27, 31, 34, 36, 41
Harvard, 79
height, 63
homogeneity, 18
homogenized, 18

I

identification, xii
implementation, 5, 67
incubation, 4
incubation period, 4
indices, 2, 10, 28
industrial, 3
industry, xii, 6, 71
infinite, 56
inhomogeneities, 77
initiation, 75, 76, 80
injury, viii

insight, 6, 25
Inspection, v
integration, 8, 65
interaction, 1, 2, 18
interface, 4
isotropic, 5, 10, 42, 61, 62
Israel, 78

K

kinematics, 11
kinetics, 76

L

lattice, xii, 4, 8, 11, 13, 14, 19, 20, 21, 25, 37, 39, 40, 46, 47, 53, 54, 56, 65, 67, 71, 72, 77, 78
lattices, 9
law, 4, 10, 14, 15, 16, 26
laws, 16
limitation, 7, 72
limitations, 65
linear, xii, 2, 4, 5, 10, 56, 61
loading, xii, 3, 5, 9, 25, 47, 56, 71, 75, 76
localization, 79
London, 79

M

magnetic, viii
manifold, 6
manifolds, 6
mask, 62
measures, 70
mesoscopic, 77
metals, 79
microstructure, 68, 77
microstructures, xii, 68, 72
mimicking, 39
modeling, xii, 4, 5, 7, 68, 72, 78

models, xi, 4, 5, 19, 25, 47, 60, 62, 67, 75
modules, 6
modulus, 2, 14, 17, 19
movement, 10

N

New York, vii, viii
normal, 2, 65
NPP, 80
nuclear, 3, 5, 6
nuclear power, 3
nuclear power plant, 3

O

observations, 5
Organometallic, iii
orientation, xi, 1, 5, 9, 19, 20, 22, 25, 26, 27, 28, 29, 30, 31, 33, 35, 37, 39, 40, 41, 42, 43, 44, 45, 46, 49, 50, 53, 54, 55, 56, 59, 62, 65, 67, 78

P

parameter, 2, 16, 41, 42
planar, xii, 5, 6, 7, 8, 9, 20, 25, 65, 67, 72, 78, 79
plants, 3
plastic, 1, 2, 3, 4, 11, 12, 13, 14, 17, 19, 27, 29, 31, 42, 44, 62, 77, 79
plastic deformation, 11, 12, 29, 79
plastic strain, 4
plasticity, xi, 4, 5, 7, 10, 16, 18, 19, 26, 28, 62, 71, 77, 79
plausibility, 8, 72
play, 3, 4
Poisson, 8, 19, 33, 35, 78
Poisson ratio, 19
polycrystalline, 1, 5, 7, 16, 18, 30, 73, 77, 78, 79
polygons, 8

poor, 8
ports, 6
power, 3, 4, 16, 26
power plant, 3
power plants, 3
prediction, 45, 79
pressure, 5, 77
probability, xii, 6, 31, 32, 33, 35, 51, 52, 54, 67, 71
probability distribution, xii
program, 79
propagation, 3, 4, 68, 72, 75, 76, 80
proportionality, 61

R

R&D, 78
radius, 33, 35
rain, xii, 71
random, xi, xii, 5, 7, 8, 9, 20, 30, 31, 33, 34, 35, 36, 37, 38, 39, 46, 53, 56, 60, 61, 62, 63, 65, 67, 71, 72, 78
randomness, 4
range, 6, 46, 56, 62
relationship, 77
relative size, 33
resolution, 30, 78
retardation, 4
rotations, xii, 65, 67, 72
Royal Society, 79

S

sapphire, 76
scaling, 7
scatter, 5, 31, 54, 55, 60, 62, 63, 65, 67
Schmid, xi, 2, 11, 15, 26, 27, 28, 29, 42, 55
selecting, 8, 45
Self, 16
SEM, 75
sensitivity, 2, 16, 39, 71
series, xii, 47, 56, 71
services, viii

shape, 4, 8, 45, 56, 61
shear, 2, 10, 11, 16, 17, 31, 35, 75
simulation, 6, 75
simulations, xii, 7, 8, 33, 63, 67, 71, 72
single crystals, 79
Slovenia, 77, 80
spatial, xii, 6, 7, 8, 65, 67, 72, 78
spin, 2, 13, 14
stainless steel, xii, 6, 71
standard deviation, 4, 60, 62, 72
steel, xii, 6, 21, 22, 23, 27, 28, 29, 30, 32, 34, 37, 71, 75, 76, 77, 78, 79
stiffness, 1, 10, 25, 26, 27, 37, 55, 60, 62
Stochastic, 78
strain, xii, 2, 4, 5, 7, 8, 10, 16, 18, 19, 26, 28, 29, 31, 32, 33, 34, 35, 36, 37, 47, 65, 72, 76, 79
strains, 4, 17, 28, 42
strength, xii, 1, 2, 9, 17, 18, 19, 71
stress, xi, 2, 4, 5, 10, 11, 14, 15, 16, 17, 18, 57, 77, 79
stress intensity factor, 4, 5, 57
stretching, 1, 13, 14

T

tension, 9
threshold, 3, 4
Titanium, iv
traction, 10, 70
transition, 4
trial, 8
twinning, 11
two-dimensional, 75

U

uniaxial tension, 9
uniform, 31, 46, 56

V

vacuum, 6

values, 3, 4, 5, 6, 22, 28, 29, 31, 32, 35, 37, 38, 42, 44, 45, 47, 54, 55, 56, 60, 61, 62, 63, 72
variability, xi, xii, 5, 71, 72
variance, 71
vector, 2, 11
velocity, 1, 2, 11, 12, 13
voids, 3
vortices, 67

W

water, 78

Y

yield, xii, 1, 2, 9, 17, 19, 71